Praise

"[Lizzie] Stark makes magic out of the simplest ingredients. Her insatiable curiosity, keen intelligence, and a literary style that is wry, elegant, and searching."

—Steve Almond, author of *All the Secrets of the World*

"Enlightening and entertaining . . . remind[s] us that even the simplest, most mundane objects may contain multitudes."

—Michael Patrick Brady, WBUR

"The book's 12 thematic chapters are dense and rich—like flan, but good." —Katherine Mangu-Ward, *Reason*

"Fun, interesting and thoroughly enjoyable, this book explores the humble egg from angles you never imagined."

—*Luxury London*

"Surprising, revealing and entertaining. After reading this delightful book, you will never look at an egg the same way again." —*BookPage*

"Science, history, art, and food come together in this quirky examination of eggs. . . . This delightful paean to the egg is equal parts fun, philosophical, educational, and irreverent."

—*Publishers Weekly*

"Stark's perky prose and awe make for entertaining reading. . . . Beyond the yolk and white, eggs are full of surprises."

—*Booklist*

"Lizzie Stark makes the egg her protagonist, tracing its biography in a dazzling probe of art, science, outer space, menus—and her body—across the globe. Imaginative and wildly original, *Egg* asks us to consider no less than the origin of everything and our role in maintaining the chicken-and-egg system."

—Marie Mutsuki Mockett, author of *American Harvest*

"Richly peppered with insight and a dash of humor, Lizzie Stark takes us on a fresh adventure to understand the humble foundation of life, the egg."

—Terry Hope Romero, coauthor of *Veganomicon*

Egg

Also by Lizzie Stark

Pandora's DNA:
Tracing the Breast Cancer Genes Through History,
Science, and One Family Tree

Leaving Mundania:
Inside the Transformative World of
Live Action Role-Playing Games

Egg

*A Dozen
Ovatures*

Lizzie Stark

W. W. NORTON & COMPANY
Independent Publishers Since 1923

For information about permission to reproduce selections from
this book, write to Permissions, W. W. Norton & Company, Inc.,
500 Fifth Avenue, New York, NY 10110

For information about special discounts for bulk purchases,
please contact W. W. Norton Special Sales at
specialsales@wwnorton.com or 800-233-4830

Manufacturing by Lakeside Book Company
Book design by Lovedog Studio
Production manager: Lauren Abbate

Names: Stark, Lizzie, author.
Title: Egg : a dozen ovatures / Lizzie Stark.
Description: First edition. | New York :
W. W. Norton & Company, [2023] |
Includes bibliographical references and index. |
Summary: "An unconventional history of the world's largest
cellular workhorse, from chickens to penguins, art to egg
crimes, and more"—Provided by publisher.
Identifiers: LCCN 2022048812 |
ISBN 9780393531503 (hardcover) |
ISBN 9780393531510 (epub)
Subjects: LCSH: Cooking (Eggs) | LCGFT: Cookbooks.
Classification: LCC TX745 .S73 2023 | DDC 641.6/75—
dc23/eng/20221109
LC record available at https://lccn.loc.gov/2022048812

ISBN: 978-1-324-07447-2 pbk.

W. W. Norton & Company, Inc.
500 Fifth Avenue, New York, N.Y. 10110
www.wwnorton.com

W. W. Norton & Company Ltd.
15 Carlisle Street, London W1D 3BS

1 2 3 4 5 6 7 8 9 0

For Roscoe

Ex ovo omnia.
(From the egg, everything.)

—William Harvey, 1651

CONTENTS

PROLOGUE

This is a book about eggs, scrambled, boiled, and cosmic; eggs in magic rituals, in the vaginas of conceptual artists, and at the center of Gold Rush gang wars. It is about crusty, obsessed Victorian gentlemen and the modern blue-collar guys who want to be like them. It is about the clown code and experiments in space. It is also about my dad, my mom, and me and our relationship with one another in and out of the kitchen.

An egg is a paradox. It is both alive and not-alive, the most precious and most worthless cell in the world: precious because it can generate new life; worthless because most efforts at creation fail, so evolution plays the odds and makes eggs plentiful. Take human ovaries. Those of us born with them start life with about one million eggs. By puberty, we're down to roughly three hundred thousand. A minuscule number of those—less than a tenth of one percent—makes it down the fallopian tubes during the fertile years. Even by the most generous measures, no human has ever used more than about a quarter of their viable eggs.* The

---

* I say "their" and not "her" because "people with ovaries" and "women" don't mean the same thing, although most people with ovaries are assigned the sex of female at birth. I have friends and relatives with ovaries who are not women, and friends and relatives without ovaries who are definitely women. I will try to preserve this distinction

world's most prolific human mother was allegedly a Russian peasant of the eighteenth century who bore sixty-nine children. History does not remember her name with certainty—it may have
been Valentina. The *Guinness Book of World Records* lauds her
merely as "the wife of Feodor Vassilyev." Let me repeat that: the
world does not know the name of the person who fit nearly sev-
enty babies through her cervix, and that—the way the patriarchy
tries to control eggs and erase mothers—is also part of this story.

Aside from its new-life potential, sometimes an egg is a fun-
gible, yet concrete commodity. One of the stranger examples of
this comes from the Philippines. Spanish conquerors spent three
centuries colonizing the islands, starting in the mid-1500s. They
brought many varieties of Catholic missionaries with them,
who built baroque churches of local materials—walls of adobe,
stone, and brick faced with plaster. The builders used animal
proteins to strengthen the concrete and plaster, including buf-
falo milk, goat blood, and the whites and shells of local duck
eggs. Forget lemons and lemonade: when colonialism gave them
egg yolks, Filipino women created iconic, yolk-heavy snacks and
desserts. These include dishes like yema, a mix of yolks, con-
densed milk, sugar, and lime zest; leche flan, a heavier version
of the classic; and a variety of cookies, Catholicism's architectural zeal reflected in rich pastry. This is a typical story featuring eggs: we start with one thing but end up with quite another,
thanks to human ingenuity.[1]

Perhaps my favorite folk tradition involving eggs is the egg
dance, a springtime custom from Germanic Europe in which
couples danced across a field strewn with eggs. If they managed
to do this three times without breaking any, then the pair could

---

throughout the book but suspect I will make some mistakes. Please
forgive me when I err.

wed without parental approval. The egg dance is a loophole, one that gets my mind wondering about potential ways around cultural conventions. Could this have been used to facilitate queer marriages? Marriages across class or national background or religion? We may never know. But the loophole that the egg dance provided—the opportunity to literally dance around the strictures of society—is a recurrent motif wherever we find eggs.

So what then is an egg? Most of the time when we say "egg" we mean a "bird egg" or, particularly, a "chicken egg," although reptiles and some mammals lay leathery ones. Species that reproduce sexually all have them: frogs lay gelatinous eggs, fish produce pearls of caviar, praying mantises create oothecas (insect egg hotels), and humans, guinea pigs, and most mammals release tiny ones. All of these may also be called ova (singular ovum)—the larger of two reproductive cells in most animals—but we call them eggs too. This book focuses on bird eggs—the hard-shelled variety—with the occasional nod to other ova.[2]

As for the proverbial question about what came first, the chicken or the egg, let's put that to rest quickly. It's the egg, man; it's always been the egg. Eggs preceded the existence of chickens by about 300 million years, when reptiles first crawled out of the primordial ooze and evolved a system for replicating away from the moisture of water: the eggshell prevents embryos from drying out. *Tyrannosaurus rex* laid eggs 65 million years ago, and chickens are their closest living relative.[3]

Biologically speaking, bird eggs are a clever little piece of tech that antiseptically contains everything a developing embryo needs in exactly the right order. Eggs originate deep inside a hen in the ovary. Birds have two ovaries, but typically only the left one is active. There, a germinal disc begins as a tiny seed that ripens. The germinal disc is the bit of DNA the bird will contribute to any potential offspring. In hens, it's large enough to see with the naked eye. Simply crack an egg into a bowl, and after a few seconds a pale spot will rise to the top of the yolk. Speaking of the yolk, it is the developing bird's food source and is attached to the germinal disc in concentric spheres, which are visible in a hard-boiled yolk sliced in half. If you look closely, you can see concentric rings of light and dark yellow. Yolk matters. Taken as a total percentage of an egg's volume, the bigger the yolk, the more developed the hatchlings will be when they emerge from the shell. Consider the northern gannet, which lays eggs with a comparatively small yolk—about 15 percent of total egg volume. Their hatchlings emerge naked, weak, and blind. The southern brown kiwi, on the other hand, lays eggs that are 70 percent yolk. Its hatchlings are "precocial," that is, they spring forth with feathers and the ability to run around and feed themselves. As for poor mama kiwi, her finished egg weighs an incredible fifth of her body weight, which would be comparable to passing a fully grown human toddler through one's cervix.

Whatever the size of the yolk, when fully ripe, it enters the oviduct, a snakelike organ with an open maw that faces the ovary. Here, in the upper part of the oviduct, sperm may fertilize the yolk. This is also where things get weird, by human standards at least. Human reproduction is a one-on-one affair. One sperm fertilizes one egg during the few days of a month that an egg is in the uterus. By comparison, chicken sex is pretty kinky. For starters, cocks don't have cocks, just cloacae, albeit ones with different abilities from the hens'. My favorite description of the cloaca—Latin for "sewer"—comes from *Ologies* podcaster Alie Ward, who declared it "like the home button on the iPhone, like if Steve Jobs designed an orifice; a multipurpose lil' boop that's good for sensual adult times, egg laying, and poo."[4] The chickens press their cloacae together, and the male ejects sperm into the female. In humans, a race of the fittest to the egg in the uterus commences, but not so in birds. A hen's body stores the sperm away inside a glandular sperm-cellar for up to two to three weeks. (Turkeys, incidentally, can hold sperm longer—up to fifteen weeks.) The hen's body passes sperm—multiple sperm—up to the ripe yolk in the oviduct. Birds, plus some sharks and amphibians, require polyspermy, a process in which several sperm must fertilize an ovum to make a viable embryo, but the reasons for this aren't precisely clear to scientists. Birds rarely lay unfertilized eggs, with one major exception: humans have bred chickens to do just that.[5]

Fertilized or not, as the yolk twists down a hen's oviduct, glands attach albumen, or egg white, to it. People talk about "egg white" like it's a single thing, but in chickens it has four layers. The innermost layer of thick white is a membrane that wraps the yolk, plus the two ropy bits known as chalazae that tether the yolk to the blunt and pointy ends of the egg and help keep it suspended in the center. The next layer of white, immediately around the yolk, is small and loose. It's the part that takes an eternity to film over when I make sunny-side-up eggs, which I usually do poorly. The two most visible parts of the white take up the most real estate in a frying pan—the tight white, which stands up tall during frying, and the final outermost layer of loose white, which spreads out thinly. The tight white also helps indicate freshness. The longer an egg spends in the fridge, the more tight white decays into loose white.

The white represents some top-shelf evolutionary engineering, as it does several jobs at once, acting as both a cushion for the developing embryo and a container for the water the embryo uses during development. From the perspective of an invading microbe, it operates like the moat around a medieval castle, creating an immense barrier for a microbe to cross. Albumen has high protein content, but when raw, the proteins are bound up in a state that microbes can't use easily. Miraculously, albumen also contains antimicrobial agents that kill and prevent bacterial growth. Some of these work best when slightly heated, say, to the temperature of a mother bird's body.[6]

After the white joins the yolk on the way down the oviduct, the egg receives its shell membrane, two meshlike layers of keratin. Anyone who has peeled a hard-boiled egg will recognize this as the layer that either sticks to the inside of the shell or the matte thing you must peel off the egg to get to the shiny white beneath. The shell membrane keeps the egg together, creates a mesh barrier to microbial entry, and shrinks soon after it is laid, which creates a crucial air pocket inside the blunt end of the egg. That air pocket beneath the shell rests next to where the chick's head will develop. Immediately before hatching, the chick punctures that membrane and uses it like a mini oxygen tank, which buys it a few hours to get its lungs going and burst out of the shell. Many species of birds produce chicks with a special egg tooth, a little horny bump at beak's end used to puncture the membrane and shell during hatching. The air pocket also provides another way for wily humans to tell how good an egg is. Whether fertilized or not, as eggs age, they lose moisture and their air pockets grow. You can test an egg's freshness by placing it in a bowl of water to see if it sinks, if the blunt end rises, or worst of all, if the whole thing floats.

With shell membrane enclosing the egg, special glands inside the hen apply the shell. This hard calcium coating protects embryos from the weight of their parents and controls gas exchange. The shell contains pores—more of them on the blunt end near the chick's head—that allow air to enter and metabolic water, the vapor living things give off, to exit. Chickens can somehow detect the oxygen concentration of the air; at higher altitudes they lay eggs with more pores to compensate for thinner oxygen.

The next-to-last layer of an egg is its pigment. Across species, birds lay eggs in a rainbow of colors: whites, creams, beiges, reds, browns, blacks, greens, and blues. So far as researchers can tell, this rainbow arises from only two pigments: porphyrin, responsible for red-brown colors, and biliverdin, responsible for blue colors. According to ornithologist Tim Birkhead's book *The Most Perfect Thing*, before an egg exits a bird, glands in the uterus work like paint guns, spraying hues onto the surface as it rotates its way to the outside. Wet pigment on a freshly laid egg can even smudge.[7]

The outermost layer of an egg is not the shell or the pigment but the waxy cuticle. In some species, it forms microspheres so small that droplets of water can't fit between them, effectively waterproofing the egg. The cuticle is an egg's first line of defense against microbes. In the United States, commercial producers powerwash the cuticle off, since barnyard detritus can stick to it, increasing salmonella risk. Since we strip eggs of their first microbial defense, we must refrigerate them to hinder the growth of any bacteria. In many other parts of the world, especially where salmonella rates are low, eggs retain their bacteria-repelling cuticle and live at room temperature on kitchen counters.

Birds stash their completed masterpieces in all sorts of places: on bare rock, in nests (their own or other birds'), and in holes. They incubate them inventively, too, with their tummies, their legs, rotting vegetation, and sun-warmed or volcano-warmed sand. Everything life requires is locked inside an ingenious, convenient, and sanitary package. The egg is a universe in a shell.

My own relationship to eggs is joyful but fraught. On the fraught side, there is my biblically bad family history of cancer and its toll on the eggs I once carried in my body. Most of the women on my mom's side have either died of breast or ovarian cancer, or preventively removed those organs to avoid cancer. I can write that simply, now, but make no mistake—my family has watched beloved relatives die horribly or suffer the lifelong effects of life-saving treatment. Knowing that you are genetically prone to develop aggressive, potentially lethal cancers at unusually young ages has created life-altering terror for many of us. For my mom and grandmother's generations, this dark cloud spurred surgeries, since you can't get cancer in organs you no longer have.

By the 1990s, researchers had discovered an inherited genetic mutation in one of the family's *BRCA1* genes that caused a strongly elevated cancer risk. Instead of waiting for cancer to strike, my generation could get ahead of it with genetic testing. When a simple blood test I took at age twenty-seven revealed that I shared my mother's *BRCA1* gene (the worst day of her life, my mother said), I planned an immediate double mastectomy, but left the ovarian surgery until I'd had a child. Still, my oncologist told me that given my family history—strong even for someone with a *BRCA1* mutation—she wanted my ovaries out before I hit forty. That is why my relationship to eggs has been fraught.

But I'm joyfully obsessed too. The December I removed my ovaries, my father and I cooked more than twenty different egg dishes in a series of experiments. We cooked eggs together for any number of reasons: I'd recently had surgery, so he wanted to show me he loved me and I wanted to show him I was OK. Plus the COVID pandemic of 2020 had made everyone stir-crazy, and my family has always used the kitchen to transport ourselves elsewhere. We "traveled" to France, where we jellied eggs in aspic; to medieval Italy, where we poached yolks in rose syrup; and to China, where we steamed them into a mind-blowingly luxurious texture. This was familiar territory for us since cooking has always been the medium of our relationship and my dad loves nothing more than arcane knowledge obsessively pursued.

During my childhood, while my mother was outside our home convalescing from cancer treatments, my dad and I bonded in the kitchen, a place that provided distraction from worry and gave a sense of control over something, anything, even if that was a boiled egg. He came up with the idea, he said, because "one of the issues with having a little girl is figuring out how to connect with her, and you know, I'm the one in our family who likes to cook the most." My real education, though, began when I turned eight and started Saturday morning swim lessons at the rec center. After each practice, my dad gave me a much-anticipated breakfast lesson. Every week we tackled a new dish. We fried, scrambled, and poached eggs. We used them to coat French toast, as our toads in the toast hole, and to raise pancakes and waffles. My dad let any mistake I made stand as a way of teaching me that my choices mattered.

Historically, I hadn't been a big egg fan, associating them with rubbery steam-plate scrambles at buffets, smelly hard-boiled offerings at Easter, or oozing fried yolks from which pancakes had to be fervently protected. When my dad was a kid, his dad used to prank him on pancake day—Saturdays—by offering him a fried egg atop his short stack. My dad called the bluff by accepting and over time developed a taste for this sacrilege. He tried the same technique on me, but I stood firm against the onslaught, mandating my egg *on the side*. To further slight him, I'd eat only the tender fried white from around the intact yolk, which I left untouched. He'd say something like, "Are you leaving that? That's the best part of the egg," and when I relinquished it, he'd pop it into his mouth whole.

At my house, one did not simply dislike a food. My father firmly believed culinary pleasure came from exposure. I didn't hate eggs; I simply hadn't acquired the taste for them yet. And though it pains me to admit it, he was right. He fed me soft-boiled eggs in little pottery cups my grandmother had made, snuck hard-boiled egg into otherwise unobjectionable potato salad, and offered me the choicest yolk-kissed bites of his pancakes. Soon I was eating my own fried yolks whole to get the oozy part over with, and seamlessly, I became a fan of most egg dishes. I still can't stomach steam-plate scrambles or browned diner omelets. I don't know how much exposure therapy is required, only that I may never reach that particular threshold.

My childhood breakfast tutorials arrived in the middle of my father's campaign to accustom me to new foods. He proposed an experiment to find the perfect soft-boiled egg. One Saturday, I arrived home from swim practice to find an absurd number of cartons—perhaps four or five—on the counter. We boiled the eggs for progressive sixty-second intervals, from the control egg (zero seconds) all the way up to a hard-boiled egg (twelve minutes), then set them neatly into a labeled carton. We knocked off their tops with a butter knife and peered at the gooey interiors. In my father's eyes, and therefore mine, the ideal soft-boiled egg has a sloppy, just-opaque white and a fully liquid yolk. We established that the ideal timing lay somewhere between three and four minutes, then ran further experiments to refine the time down to the second.

The following Monday, I flew down the stairs to boil the perfect quick-breakfast dish with my father. A few minutes later, I placed the ideal egg onto my toast and whacked the broad side sharply with a knife. Imagine my disappointed little face when a gush of practically clear egg snot came running out.

And that, dear reader, is how I learned an important lesson about experimental conditions. We'd used room-temperature eggs in our weekend project, but on a busy Monday morning we grabbed them cold from the fridge. Lesson one: eggs cook fast, so starting temperature matters.

Our culinary experiments have continued ever since. Though not all of them were egg focused, we always had fun. As we cooked, we'd discuss life stuff—my friends, interests, and hobbies. As I grew out of the narcissism of childhood, I learned more about my dad: that his lawyerly work combined research and writing, that he loved the trumpet almost as much as cooking and could digress extensively on the differences among trumpet mouthpieces, and which collection of Shakespeare on tape reigned supreme. Even now, cooking time is when my dad gives me fatherly advice, praises my son, and tries to get me interested in new, and always mythical, cooking experiments. The kitchen is our clubhouse, the experiments the price of entry.

I never needed that clubhouse more than in the fall of 2020, when I carved my own personal supply of eggs out of my body to avoid a fate passed down in my mother's family. The rest of the world joined me in medical limbo as COVID spread internationally. Those of us who had enough time and money retreated into the shell of homelife and headed straight for the kitchen.

This book examines more than eggs in the kitchen, though. This is a book about what one creative species can do with the world's largest cellular workhorse and what that says about us and about power. It is about cosmic loopholes, the ingenuity of the oppressed, and the nature of obsession. It is an attempt to understand what I gave up when my ovaries came out. It's my attempt to reflect on the meaning of an object that carries deep symbolism and heavy metabolic importance. And it is also a means of traveling through stories of eggs to other times and places.

So please, come sit in my kitchen. Relax. Open the carton, and take a look at a dozen perfect and perfectly protean eggs.

Egg

1

COSMIC EGG

An egg is perhaps the world's most primal symbol. It appears in a staggering number of mythic traditions, from India to Finland, Mali to Japan, Tahiti to Greece, and far beyond. In these stories, the egg is the beginning of all things, the vessel that cracks open and releases the world. As a motif, it's so common that scholars have given it a name: the World or Cosmic Egg.

These origin myths take a few different forms. Often, the egg cracks open and lets out the world, as in the Finnish myth from the *Kalevala*. In the nineteenth century, linguist Elias Lönnrot traveled the country, recording ancient oral traditions of stories and poems. The result, published in 1835, was the *Kalevala*, widely considered the Finnish national epic.

As with most origin myths, the *Kalevala* begins *in medias res*, provoking chicken-or-egg questions, with lonely primordial waters and a beautiful duck flying over them searching for a place to nest. The Mother of the Waters, the personification of primordial ooze, lifts a leg above the waves for the duck to land on. The bird lays six eggs, five of gold and one of iron. She incubates them for three days. When the heat of the bird and her eggs becomes unbearable, the Mother of Waters jostles her leg for relief. The eggs roll down her leg and splinter. As one translation of the story goes,

> And the eggs fall into ocean
> Dash in pieces on the bottom
> Of the deep and boundless waters
> In the sand they do not perish . . .
> But transformed, in wondrous beauty
> All the fragments come together.[1]

In the fertile waters, the broken eggs merge. The shell fragments fuse into two large pieces, an upper one that forms heaven's arch and a lower one that becomes the Earth. The yolk transforms into the Sun, the whites become the Moon, the "dark part" becomes the clouds, and the shells' speckling turns into the Milky Way. The Mother of Waters, delighted by the creation, shapes the contours of the world with her stomping and swimming. Soon she gives birth to an old wise man, Väinämöinen, who floats on the waters for nine years and then finishes the act of creation by sowing the lands with plants. His male force completes the act that all that chaotic feminine energy began, which is also a recurring motif in cosmic egg stories.[2]

The Dogon people of Mali and Burkina Faso live on and around the Bandiagara plateau, an escarpment that runs through Mali, close to the two countries' borders. (I had the good fortune to hike the escarpment and see both abandoned and populated cliff villages in 2000, before civil war made travel in the region impossible.) Their creation myth begins with an egg, fertilized by Amma, the supreme being, who uses he/him pronouns but nevertheless contains both male and female elements. The universe shakes up Amma's egg seven times. The stirring divides the interior of the egg into two placentas, each containing a male-female pair of twins. Like Amma, each individual twin has an identified sex of male or female but is also androgynous, containing elements of femininity and masculinity. The twin Yurugu breaks out of the egg early, and the piece of placenta he dislodges becomes the Earth. He returns to the egg to find his twin sister, but she is not there: she has been placed in the egg sac with the other pair of twins. Yurugu tries to have sex with his own placenta, but nothing happens. So Amma sends down the rest of the twins to copulate, and they create humanity. Here, humanity, with all its divisions, comes from a single cosmic egg and from gods that are simultaneously both singly and doubly sexed.[3]

Meanwhile, in a Tahitian creation myth, the god Ta'aora does not give birth to a shell but lives inside one that revolves in space. When it cracks open, he springs out, calling forth the world. He makes the shell into the dome of the sky; he gives his spine and ribs to make the mountains and his flesh for the strength of the Earth. His fingernails and toenails become fish scales, and his blood becomes rain and rainbows. He also gives every living thing on Earth its own shell, saying, "Man's shell is woman because it is by her that he comes into the world; and woman's shell is woman because she was born of a woman." Here, women are figured as a kind of human cosmic egg. We are ourselves but also the vehicles for future generations. That uneasy tension has played out in an abundance of different ways across the history of humanity. We are considered to be half agent, half object; half person, half advanced ICU unit.[4]

The cosmic egg motif pervades the world's oldest texts, from India's *Chandogya Upanishad* and *Rig Veda*, to the *Nihongi*, Japan's second-oldest book of history. It makes appearances in Slavic and Chinese myth cycles alike. So primal is the egg as a symbol that in ancient Egypt, which also has a cosmic egg myth, the glyph for "egg" is part of the name of every goddess and appears in symbols for a range of linked concepts, including those for humanity, fertility, phallus, pregnant woman, womb, birth, and even coffin, since a coffin is a shell for those awaiting rebirth.[5]

These ancient myths share themes. The egg represents fertile, generative, and often feminine chaos. It either exists on its own at the start of the story or arrives courtesy of a female or androgynous creator god. Parts of the egg often become the world: sun/yolk and shell/heaven associations are widespread. A creator, male or androgynous, often springs forth from the shell, bringing order to the chaos or having sex with themself to beget the world.

Interestingly, almost no one in my life seemed to know about the cosmic egg. I found this strange since my friend-circles contain many story obsessives: beleaguered adjunct professors, writers, interactive theater designers, budding psychologists, and others with rich interior lives. When I was a kid, world myths obsessed me—for two or three years they were all I read—and yet no memory of a cosmic egg has remained. That forgetfulness, that sub rosa quality of the cosmic egg echoes the status of real eggs. Eggs are everywhere but simultaneously invisible, as women's work, or at least the work of female bodies, often is. Part of this project is to render that labor visible.

Since eggs are a primal, mystical, ancient symbol, it's no surprise that they show up in religious and magical traditions. For example, the ancient Romans lived by auguries, and birds, which occupied the airy realms of the gods, made natural messengers between divine and earthly beings. This presented a problem for soothsayers: staring at the sky until an interpretable bird flock flew by was incredibly inconvenient. A tame flock of sacred chickens, on the other hand, could weigh in at a moment's notice.

Upon request, the *pullarius*—the sacred-chicken priest—would fling food into the chickens' pens and watch what happened. If the birds ate, that was good; if they didn't, that was bad; and if they fled, well, that was the worst. Of course, *pullarii* could easily manipulate the signs if it was politically expedient: starve the chickens for a few days, and the omens would point in your favor.[6]

The Romans kept sacred chickens everywhere they might be useful—in temples, with the army, and on ships. One memorable consultation occurred during the First Punic War, a series of battles fought between Rome and Carthage over Sicily and the north coast of Africa as Rome tried to expand southward. At the time, two consuls ruled Rome by democratic election. One of them, Publius Claudius Pulcher, was a member of Rome's old and influential Claudii family, and his behavior toward the sacred chickens led to his downfall. Before his fleet of 123 ships launched its surprise attack on the harbor of Drepana in 249 BCE, he decided to consult the oracles. When the chickens refused to eat, he had them thrown overboard, saying, "If they won't eat, let them drink." Many believed his impiety had ill consequences. The chickens may have drowned, but Pulcher lost 93 ships and tens of thousands of men. Home in Rome, he was charged with treason and fined 120,000 asses, an *as*, of course, being a bronze Roman coin.[7]

Where there is chicken divination, surely oomancy follows. When Livia Drusilla became pregnant by her husband, Tiberius Claudius Nero, a politician in ancient Rome, she incubated an egg between her breasts in hopes it would foretell the birth of a son. It worked. She hatched a rooster, and in 42 BCE gave birth to the future emperor Tiberius Caesar Augustus. In ancient Egypt, meanwhile, tomb walls and clay tablets written in hieroglyphs reveal spells incorporating eggs; sailors could pray to a clay egg for safety at sea.[8]

Wondering whether eggs might still be a part of magical traditions, I turned to my high-school friend Kristen Sollée, a second-generation witch and author of *Witch Hunt: A Traveler's Guide to the Power and Persecution of the Witch*. She is a year younger than me and undeniably cooler. In high school, we held down the second alto section of the *a cappella* group. Although we hadn't talked in a few years, she agreed to help with my research and sent me some history information and spells, including information about John Hale. On matters of witchcraft, the reverend of Salem, Massachusetts, was a flip-flopper. He helped prosecute many women during the witch trials on a "these-women-are-evil" platform before changing to the more moderate belief that "these women are possessed by demons," as set forth in his posthumously published work *A Modest Enquiry into the Nature of Witchcraft* (1697). In it, he discusses the case of a poor teenager who tampered with the "Devil's tools"—you know, a glass and an egg. She was trying to find "her future Husbands calling; till there came up a Coffin, that is, a Spectre in likeness of a Coffin. And she was afterward followed with diabolic molestation to her death; and so dyed a single person." He doesn't specify exactly how she used the glass and the egg—I suppose he didn't want others to try it—but I want better proof for "the devil killed her" than his say-so. As a former teen girl, I can tell you that trying to predict the identity of one's future spouse is a popular activity. In my day, we tried cards, ouija boards, flower petals, apple cores, the pop-top on soda cans, and much more. Over twenty years later, not one of us has been hounded to her death by a demon. Maybe it's simply that we left our parents' eggs unmolested in the fridge. Nor do we consider dying single as a fate worse than death, as Hale seemed to.[9]

Either way, reading Reverend Misogyny got my hackles up, so I felt inclined to tempt fate by trying an egg cleansing, also called a *limpia*, using instructions Kristen sent me from the Hoodwitch website. I'm not a superstitious person per se. I'm married to a bioinformatician, after all, a man who thinks that God can show us his receipts just like everyone else. As a story lover, I like the idea of spirits being real, although I can't bring myself to believe in something unproven and unprovable. What I do believe in, though, is the power of ritual to change mental states, a sort of tonic for the parts of our brains we cannot actively control. *Limpia* comes out of Mexican and Mesoamerican magical tradition (though similar practices exist elsewhere), and the word means "cleansing" or "purification." It's often shorthand for an egg spell, Kristen wrote, and is designed to dispel negative energy and mental blockages.[10]

After the 2020 elections, we had a week of torturous limbo before the winner was declared. It seemed as good a time as any to try a ritual. As this was during the pandemic when we were in quarantine, I did not have anyone from the tradition guiding me through the practice, only the sparse Hoodwitch text Kristen sent. I did it as well as I could, washing a raw egg in salt water and holding it in my hand while I meditated on what I wanted to cleanse my space of—bad election vibes that blocked me from writing. Then I rolled the egg around my kitchen, couch, and office. This proved harder than it sounds since eggs don't roll in a straight line. Eventually, I got the hang of guiding it between my thumb and forefinger. The damp shell picked up dust from my kitchen counter and the place on the couch I sit when I am pretending to do work. In my office, it typed "fsaaazk'][jhf" on the keyboard. According to the spell's premise, the egg absorbs negative energies that one discards along with it. I cracked the egg into an empty bowl with a satisfying rap and examined it. This egg had an unusual feature—double the usual number of chalazae. Typically, an egg has two ropy bits of egg white holding the yolk suspended within the shell. My egg had four. No wonder I was feeling blocked. I tossed the obstructions into the garbage. Whatever you think of magic, this was exemplary experience design. I had to ponder my desired outcome, visit the places where I wanted to change my behavior, interact with them in a new way, and then complete the cycle by breaking and throwing out my symbolic little stone. I don't know whether it worked, but it sure felt good.[11]

If we crack the cosmic egg a little further, many traditions—of science, religion, and food—fall out. The magic of eggs and magical thinking around them influenced the proto-science of alchemy, which I learned from my uncle Alan Rocke. He is professor emeritus of the history of science at Case Western Reserve University, where he also taught food history. He's tall, slender, bespectacled, and mustached, with a kind and gentle face. It is a mistake to play him in Scrabble. He is also one of the world's leading experts in the admittedly niche field of the history of organic chemistry. He explained to me that alchemists viewed eggs as the source of life, and since the goal of alchemy was a "universal medicine that conferred immortal life and perfect health," eggs proved an appealing ingredient. On a visual level, the egg yolk is the color of gold, which made it, quoting Uncle Alan, "in some way divine," according to the alchemists. They stirred up their experiments inside egg-shaped flasks, which represented the mythical vessel that fused the universe into existence.[12]

Other modern scientists have found the idea of a cosmic egg inspiring. Space historian Dr. Jordan Bimm, a postdoc at the University of Chicago, wrote me that astrobiologists love to think about the origin of life on Earth and that, as he put it, "One famous hypothesis is 'panspermia,' the idea that life arrived on Earth after traveling through space on meteors and comets. This metaphor figures the Earth as a giant egg!" The jury is still out on whether the theory is correct.[13]

Crack the cosmic egg, and you can make a mystical omelet as well. Eggs are ubiquitous. They thrive on every continent and within an astonishing array of environments, from equatorial jungles to the darkest day of Antarctic winter. They are a primal food; humans have eaten eggs since the literal dawn of human time. Archaeologists have found roasted egg fragments in Australia dating back some fifty thousand years. Those eggs belonged to emus and the flightless *Genyornis newtoni*, a seven-foot-tall, five-hundred-pound behemoth that looks like the results of a threesome between an ostrich, a goose, and a dinosaur. Those beasts laid eggs the size of cantaloupes before humans ate them to extinction. I suspect one of the reasons eggs are such a popular foodstuff is that—so far as I can tell—nearly all animal eggs are safe to eat. Ninety-nine percent of the world's one million to two million known species of animals reproduce through eggs. Remarkably, only a handful of species lay poisonous ones, so be warned: don't eat the roe of the saltwater cabezon or of any of the seven species of freshwater gar. Likewise, give the eggs of several bird species from New Guinea a miss, including those of the hooded pitohui and blue-capped ifrit, which coat their eggs in toxic preen oil. Simply put, eggs are universally available and overwhelmingly safe to eat: of course ancient hungry humans used them for lunch. Human nature also made it inevitable that we would create numerous customs and taboos around eggs on the plate.[14]

For starters, eggs were once a seasonal food. The reproductive cycles of birds, including chickens, are sensitive to light, so most birds lay their eggs during the spring and summer. Eating an egg out of season meant eating an old egg, one preserved in a coat of mineral oil or wax, buried in sawdust, hard boiled and then pickled, or buried in salt, quick lime, or lye-laced clay—that is, if you ate eggs at all. According to the *Cambridge World History of Food*, in many parts of the world people avoided eating eggs. "In part this was because eggs were regarded as filthy food (the product of the semen of a cock), in part because of food taboos (such as those that regulated the diet of pregnant women and their youngsters), and often because it was believed to be wasteful to eat the egg instead of waiting for the chicken." Eggs combine all sorts of interesting cultural values, like thrift—saving the eggs to eat the chickens; taboos against sex, making eggs too scandalous for the dinner plate; and the ancient preoccupation with telling pregnant women what to eat. By the time of my grandmother, pregnant women were supposed to eat eggs. She bought a flat of them while carrying her eldest. They sat out at room temperature in the apartment she shared with my grandfather, who was off at World War II. She aimed for an egg a day, hard boiled, but by the end of the month she had to eat around black spots in them. My dad and his siblings, fluent in sarcasm, say this explains a lot about their sister.[15]

At any rate, what is religion but a series of taboos and dictums? Eat this, not that. Praise this god, not that one. But these taboos have changed over time. Japan, for instance, consistently ranks among the top three countries in the world for per capita egg consumption, astonishing if you consider its eggless history. From the seventh to ninth centuries, emperors repeatedly banned eating animals, including eggs, based on Buddhist prohibitions against killing, as well as on the indigenous religion of Shinto, which viewed chickens as sacred messengers from the gods and heralds of the dawn. According to a paper by Naomichi Ishige, food historian and director of the National Museum of Ethnology in Osaka, most Japanese people avoided eating eggs until the mid-1500s, when Europeans and their Christian missionaries made contact. As some locals converted and began to change their diets, the Buddhist monks launched a public relations campaign that included spreading rumors that Christians were cannibals (not so far-fetched, given the rhetoric around communion) who drank the blood of children. Fearful of foreign influence, Japan began its Sakoku policy of isolation in 1603, which effectively stranded their population at home. Locals couldn't voyage out, and foreigners couldn't do anything but make port calls. The policy also renewed bans on using horses and cows for meat. An occasional chicken was still OK, though, and egg consumption began to grow. Eggs didn't become big in Japan until Sakoku ended with the Meiji Restoration in 1868. To compete with the United States and Europe on the world stage, Japanese intellectuals believed they needed a robust standing army, one made as strong as the western armies were made by eating meat and milk. As Ishige put it, the government promoted egg consumption "as it was thought that their inclusion in the diet might also lead to improved physique and strength." Perhaps the only thing that can crack a religious prohibition is an economic or military imperative.[16]

In Judaism, meanwhile, eggs have always occupied a middle ground between meat and not-meat. Conveniently for people who keep kosher, eggs are considered *pareve*, that is, a neutral food that swings both ways: you can eat them with meat as well as dairy. An iconic part of the springtime Passover Seder plate, hard-boiled eggs represent new life. The Talmud, the primary text of Jewish law and tradition, brims with details about when and how exactly one can eat eggs on the Sabbath when cooking is prohibited. In her paper "Eggs in the Talmud," Susan Weingarten, historian of Jewish food, explains that "eggs should not be put to roast before the Sabbath unless it is clear that the roasting can be completed before the Sabbath begins." These prohibitions gave us the Sephardic dish huevos haminados, eggs hard-boiled on the stove or in the oven, low and slow on their own or in stew for many hours.[17]

My dad and I, of course, have boiled eggs with the best of them. We've done one-, two-, three-, five-, and eleven-minute eggs. But an egg boiled up to a dozen hours, beyond our previous record of two hours (for Chinese tea eggs, steeped in soy and tea)? That we had to try. After a few hours of research, we selected two recipes, one from cookbook author Joan Nathan, the doyenne of Jewish cooking, and the other by a food website user known as "Sephardi Kitchen." We made them in tandem. Nathan's recipe involved braising eggs in water, sliced onions, and oil; peeling them halfway through and continuing to simmer them; then reducing the liquid and using it to sauté spinach. Sephardi Kitchen's version did not shell the eggs but cracked them halfway through the process, simmering them in a mixture that included coffee and papery onion skins (for color) as well as garlic and vinegar. Seven or so hours later, we fed our spouses the results—crumbly yolks and firm cream-colored whites that had a mild but distinct barnyard note. We favored Nathan's version, because the eggs had a more intense flavor—likely because we simmered them peeled rather than merely cracked. We also preferred them because eggs cooked upward of two hours are dry dry dry, and her recipe had juicy spinach to mitigate that. As my dad put it in the debrief he posted to Facebook, where he chronicles many a cooking project, "The differences from simple hard-boiled eggs were fairly subtle." They weren't bad, but absent a religious or sentimental reason, we probably won't make them again.[18]

While the Talmud classifies eggs as between meat and milk, medieval Christians started out firmly in the "meat" camp. Medieval Christians fasted a lot—Wednesdays, Fridays, Advent,

Lent, and during other periods. On fasting days, which num-
bered roughly half the year, they couldn't eat meat, only fish.
This prohibition had a predictable effect: fish-fatigued believers,
including the clergy, began an avid search for loopholes. Rhetor-
ical gymnastics classed foods like puffins and beaver tail as fish.
After all, puffins mostly float on the sea and are thus fish and not
birds, while beaver tails spend most of their time in the water
and don't taste all that good—and isn't that the point of fasting?
People desperate for eggs even concocted vegan versions out of
almond milk, straining it to fit into an eggshell and then roasting
it, dyeing the "yolk" yellow with saffron. Eventually, during the
Renaissance, Catholic restrictions loosened to permit eggs and
dairy on fasting days.[19]

As these religious loopholes show, an egg is a contradiction.
It is meat and not-meat; it is alive but not quite living; and it has
a ubiquitous but not well-known (at least in the United States)
mythic origin story. It is an object of magic, mysticism, religious
fervor, of utmost seriousness as well as humor.

It is also at the center of my personal mythology, forged in my
father's kitchen. My grandpa, for whom my son is named, loved
kitchen experimentation. As my dad put it, he was "always interested in food that had some sort of—mythological isn't quite
the right word, but some adventurous aspect to it," whether
that meant roasting a suckling pig or trying out fancy French
omelets. My father inherited that love of epic, mythological food
and passed it on to me. He cemented his paternal bond with me
during those Saturday-morning breakfast lessons in boiled eggs.
As I got older, we graduated to eggs Benedict with yolky hollandaise, soft French scrambles, and buttery omelets.

Homey yet versatile, he couldn't have picked a more epic ingredient for our shared bonding. Eggs are cheap and cook quickly
but remain edible even when abused. In other words, they are

an ideal way for a father to teach his daughter any number of useful lessons, and not just about how to cook breakfast. Cooking eggs with my dad taught me about my own agency. At the stove, I shook the omelet pan, and my skill and technique had a marked effect on the finished dish. By making many omelets in a row, I learned the power of rapid prototyping—changing up small variables, eating my own mistakes, or deciding to start again or fail them forward into a slightly tough but still pretty good scramble. Cooking an egg is like jazz: once you master the basics, you can combine them in any way you like, and it'll come out all right. On a personal level, that is my own cosmic egg myth, cracking shells freestyle in the kitchen with my dad and learning that at the stove I control a dish's destiny, using heat to form the chaos into a contradiction, an orderly, panned, but not-quite-planned golden crescent on a plate.

EGG HUNT

Major Charles Bendire spent his military career stationed at remote outposts on the US frontier, where he helped his adopted country live out its genocidal manifest destiny. To fight the boredom of a life away from civilization, he got into birding and egging. In 1872, while posted in Arizona, he spotted a zone-tailed hawk during a patrol and climbed a tall tree to reach its nest. As soon as his hand touched the rare egg, an Apache scout appeared and began shooting at him. Resolved to save his find, Bendire did the only thing he could think of—popped it into his mouth for safekeeping, shimmied down the tree, and lit home on his horse. Back at the camp, he discovered a problem: all that excitement had locked up his jaw muscles, and he couldn't unclench them enough to spit the egg out. A different man might have broken the egg to remove it, but Major Bendire had a collector's passion. He ordered his men to extract it, whole, under threat of court martial. So they pried his jaw open and got it out, breaking only one of his teeth in the process. Bendire rode his egg obsession all the way to the Smithsonian National Museum of Natural History, where he became the honorary curator of eggs; his collection of eight thousand calcified shells became the basis of its oological holdings.[1]

People who can't get enough eggs turn into weirdos. Perhaps

that's a bit rich coming from me, but it's true. Men—it's always men—have gone to the literal ends of the Earth, all for the sake of an eggshell. In the nineteenth century, oology, the study of eggs, proved a popular hobby for boys. More wholesome than collecting cigarette cards, eggshells were the Pokémon of the Victorian world, supporting multiple magazines, including *Young Oologist*, for children. With tens of thousands of species needed to truly collect 'em all, it's no wonder certain boys grew into obsessed men. Many kept up their collections for aesthetic reasons. Nature paints eggs in delicate colors from blue to green, brown to red to white, shiny or matte, a solid color field or speckled and squiggled. Eggs are beautiful. Modern research suggests that our brains may be hard-wired to find them so. Psychological research from 2010 and 2013 used MRIs to measure human brain activity while the subjects looked at art and architecture. The studies found that curvilinear and ovaloid shapes—a fair description of eggs—excite the brain with pleasure. Back in the Victorian era, some collectors specialized in eggs from certain parts of the world, while for others, the gold standard meant robbing a single mama bird of every egg she laid over her lifetime. Guillemots, a family of cliff-dwelling seabirds who lay pear-shaped eggs in a range of colors, with a pattern of brown squiggles unique to each mother, were particular targets.[2]

Collecting eggs required a veritable Swiss Army knife of skills. One had to be a keen observer, noticing birds of a given species and then tracking them to their nests. In springtime, collectors had only the span of a few days to nip in for their booty. If they waited too long to collect a clutch, they wouldn't be blowing out a yolk but dissolving an embryo with acid. Oology also required agility and strength. One had to scare off the parent birds—the danger in trifling with, say, female raptors no doubt provided some of the thrill—and scale a tree or rappel down a cliff.

Plenty of eggers plunged to their deaths during the spring season. Frances J. Britwell couldn't stay away from his passion, even on his honeymoon in the Sierra Nevada. He climbed a tall pine in pursuit of a nest but slipped and strangled to death in his own climbing ropes while his new bride watched. Collectors also perished from gun accidents: the surest way to identify an egg, of course, involved shooting its parents and collecting them as well. Gunshots could kill a collector directly or cause death from infection, as was the case with the University of Munich's Dr. Johann Wagler, who accidentally shot himself in the arm while collecting. The buckshot wounds became infected, and he died of blood poisoning at age thirty-two. Assuming the collectors survived their shooting expeditions, they then preserved the bird skins for study. Picture taxidermied birds without the stuffing or lifelike posing, lying in rows, feet up, in filing-cabinet drawers. The act of preservation also presented a danger: collectors used arsenic during the process, and handling skin collections frequently posed a risk of poisoning. It could leave collectors with diminished appetites and tongue ulcers and may have caused or contributed to actual deaths. Collecting eggs on steep sea cliffs carried a risk of plunging into the ocean and dying on impact or by drowning. European collectors in distant locales grew ill from exotic fevers or ran afoul of angry locals. Dr. Heinrich Macklow, for example, collected with a group of six naturalists. Over five years, half of them died of yellow fever. In 1832, Chinese workers in Eastern Java burned down the house where Macklow was staying; he survived the fire but died by the spear a few days later, after joining a posse bent on revenge. I rather imagine the locals thinking, "What is it with these ridiculous white guys, running around and killing all the local fauna?"[3]

Once oologists collected the eggs, they had a craft project ahead of them. They used a hand drill to cut a small hole on the broad side of the egg, inserted a blowpipe smaller than the hole, and then used air to force the contents out. This isn't easy, as I learned in a few egg-blowing experiments of my own. To eject the egg's innards requires considerably more force than blowing up a very thick rubber balloon. I found myself lightheaded and exhausted after getting perhaps a quarter of the contents out. The process is also disgusting. Egg innards go down the metal tube and drip over your hands and onto the table or bowl below. After emptying the shell, collectors used a syringe to fill it half full of water, shaking it to rinse out the remaining egg stuff, then repeating the process until the water ran clear. They let the egg rest, blowhole down, on absorbent cornmeal to drain and dry. Thanks to Major Bendire's work popularizing methods for egg collection while he worked at the Smithsonian, most collectors marked their finished eggs in soft pencil near the blowholes, with the date of collection, number of eggs in the set, and so on. Many collectors also kept detailed notes of their finds, which they recorded in journals or on small cards kept with the specimens. The finished eggs were wrapped in cotton for transport and often displayed in partitioned cigar boxes with custom glass covers.[4]

But before any display is possible, you need to find the eggs. And not every egging trip revolved around the mere acquisition of shells; sometimes collectors wanted the embryos too. One such adventure was, hands down, the world's strangest. It began on the Ross Ice Shelf in Antarctica on nearly the darkest day—June 27—of 1911, when three explorers hiked into the endless night on something of a side quest for emperor penguin embryos in hopes of settling a Darwinian debate. We know of the trek from one of the explorers, Apsley Cherry-Garrard, called "Cherry," who was part of Captain Robert Falcon Scott's doomed *Terra Nova* expedition to reach the South Pole. Cherry was a member of the idle rich who loved to travel but usually did so on cruise ships, and the crew nicknamed him "Cheery" for his disposition. He wormed his way onto the expedition thanks to personal connections to its science officer, Dr. Edward Wilson, and a sizable donation he made. Captain Scott initially rejected Cherry's application to join the party but was so impressed that Cherry didn't then renege on his financial obligation that he reversed course and allowed the young man to join them.[5]

Cherry wrote a travelogue of his experience titled *The Worst Journey in the World*. Certainly, the expedition ended very badly, with the Norwegians beating the British to the South Pole and everyone from the English polar party slowly freezing to death on the return trip just a few days' journey from where Cherry waited with dogs and supplies. The dying men retained enough fortitude to write stiff-upper-lip letters to their survivors that would sustain the sermons of British rectors for generations. And yet the "worst journey" Cherry wrote about was not the doomed polar part of the expedition. It was the trip to Cape Crozier for eggs.

Another member of the penguin excursion was Henry "Birdie" Bowers, so nicknamed for his beak of a nose. He was a red-headed Scottish seaman of robust constitution, and, as author and Antarctic historian David Crane put it, he had "a rich portfolio of prejudices" against a variety of nationalities and religions. The final member of the trio and commander of this journey was Dr. Wilson, a close friend of Captain Scott's. He and Scott had bonded on a previous expedition to Antarctica, on the *Discovery*, over their shared dislike of its leader, a man they managed to save from certain death after he collapsed, coughing blood, on a long foot journey over the snow. The *Discovery* spent two years jammed in ice during that expedition. During that time, Dr. Wilson first discovered the Cape Crozier emperor penguin colony and took an interest in the birds, visiting them several times and studying them. From his visits, which did not end with possession of penguin eggs, he concluded that the birds must lay their eggs during the height of winter.[6]

As Captain Scott prepared for his summer try at the South Pole by laying supply depots, he let three of his men traipse off on a 120-mile round-trip walk in extreme winter weather. A central question of the journey, historian Crane wrote, was "why Scott should ever have allowed it to take place." And yet he permitted it because he and Dr. Wilson had fought death together, and Dr. Wilson really wanted those penguin eggs.[7]

Wilson's egg mania had to do with an arcane and disputed evolutionary argument of the day. Basically, scientists noticed that embryos of different species resembled each other. The theory is summed up as "ontogeny recapitulates phylogeny," that is, the physical development of any embryo reflects its past evolutionary history. For reasons I haven't been able to ascertain, people believed that emperor penguins were an ancient species and thus that their developing chicks might show evidence of a link missing from the evolutionary record—the link between birds and dinosaurs. Wilson wanted to be the man who found the proof.

Cherry summed up the trip as an experience that "beggared our language: no words could express its horror." His memoir paints a pretty good picture of hell, though. Team Penguin pulled 750 pounds of gear on foot on two sledges. The weather conditions averaged –50°C (–58°F), cold enough that the pus inside their frostbitten blisters froze. Within a few days, the water vapor their bodies gave off had impregnated their clothes and other gear. It then froze. The accumulated ice nearly doubled the weight of their tent and bags by the end of the trip. Each morning, they fixed their heads into the day's position because their balaclavas froze stiff within seconds of hitting the open air. At the day's miserable end, they spent forty-five minutes melting their way into icy sleeping bags, to lie shivering for a few hours before rising to walk again. At times, the weather got so cold that the friction from the sled runners couldn't melt the snow. The sleds no longer slid, and had to be muscled through what felt like fine sand. The men would work together to pull a single sledge one mile, then backtrack to find the other and pull it forward—three miles of walking for every one advanced toward their goal. No wonder that Cherry spent much of the journey fondly imagining the sweet sleep of death at the bottom of a crevasse. Under his breath he chanted the mantra, "You've got it in the neck—stick it—stick it—you've got it in the neck," over and over again.[8]

After nineteen days of frozen hell, the team finally arrived at Cape Crozier and set up camp. They made a dangerous descent down to the penguin colony, where they discovered only a few hundred nesting birds, a far cry from the thousands Wilson had seen on his previous trip. Still, they managed to snatch five eggs. Cherry broke two on the way back due to his nearsightedness but still, sweet victory! At camp Wilson cut square windows into the remaining three eggs to reveal their embryos, then pickled them for further study. They could always find more the next day.

Except a hurricane-force blizzard struck that night. The tent where they had secured their supplies blew away, as did the covering of the rock shelter they had built. They spent the next two days in their sleeping bags under drifting snow, singing hymns to confirm that everyone still lived. When the storm cleared, they found a miracle: their tent had not blown out to sea but was only five hundred yards away. With diminished gear, they decided to cut their losses and head home. The return journey, made shorter by more cooperative weather, had its own challenges, as Cherry was about to discover one of the then-unknown side effects of prolonged exposure to extreme cold. I'll let him tell it. "We stood panting with our backs against the mountainous mass of frozen gear which was our load. There was no wind, at any rate no more than light airs. Our breath crackled as it froze. There was no unnecessary conversation: I don't know why our tongues never got frozen, but all my teeth, the nerves of which had been killed, split into pieces. We had been going perhaps three hours since lunch."[9]

Ghostly versions of Cherry, Wilson, and Birdie finally returned to base camp, where their colleagues cut the reins of the sledges out of their hands. Captain Scott described them as "more weather-worn than anyone I had yet seen." But they had the eggs they came for.[10]

The polar party set out five months later, in November 1911, and as the group laid various supply depots, Scott sent waves of men to return to base camp. Wilson and Birdie would be part of Scott's final polar party, along with two other men. Cherry was in the second wave of men to be sent back to base camp. Through a series of misadventures, he ended up bringing a dog team to one of the depots to meet Scott's party on their return from the pole. Unbeknownst to him, three days' journey south, the final remains of the party—Scott, Wilson, and Birdie—were slowly starving to death. Cherry joined the group that would recover their bodies eight months later, along with artifacts of the polar trip, including biological samples and letters.

In 1913, after returning from Antarctica and with that pathos weighing on him, Cherry visited the British Natural History Museum to donate the penguin eggs. The curator, whom Cherry described as "extraordinarily offensive, even for an official man of science" took the eggs "without a word of thanks" and told him to get lost. The eggs, alas, proved nearly nothing. The first researcher who received them used them to make slides but died before he could study them. Eventually, the slides found their way to zoologist C. W. Parsons, who published a paper on them in 1934. By then debates about whether embryos summoned their evolutionary ancestors were long over: they didn't, ontogeny does not recapitulate phylogeny, and thus penguin embryos couldn't settle the bird/dino question after all. As a final insult, Parsons concluded that the eggs "did not greatly add to our understanding of penguin embryology."[11]

The Cape Crozier penguin eggs present something of a parable for the science and hobby of oology. All these eggs were supposed to add up to something. In *The Most Perfect Thing: Inside (and Outside) a Bird's Egg*, British ornithologist Tim Birkhead states that eggs were supposed to "provide material from which the natural order can be deduced," and yet, for this purpose, Birkhead concludes, "eggs proved to be almost entirely useless." In the end, we have the Freudian image of obsessed men surrounded by cases bearing thousands upon thousands of the fruits of female bodies.[12]

As a discipline, oology is basically deceased. It had been the gateway drug to harder types of ornithology: egging and skin collection provided an accessible way to get into a serious bird obsession. But once high-quality binoculars and cameras became affordable, birders no longer required trophies to make identifications. Around the same time, two Boston socialites—Harriet Hemenway and Minna Hall—crusaded to end the widespread slaughter of birds for ladies' hats, spawning a transnational movement that resulted in the passage of several laws limiting bird murder in the United States and United Kingdom. As for the eggs themselves? These data-rich collections languished in museums, mostly unused, save for a few researchers who made occasionally shocking finds. Most famously, in the 1960s, predator bird populations fell because eggshells could no longer withstand the weight of the nesting female birds. They were crushing their eggs before they could hatch. Joseph Hickey and Daniel Anderson had the bright idea of comparing contemporary egg-shell thickness in areas with and without DDT exposure to those from old egg collections. The result, published in *Science* in 1968 and later confirmed by a flurry of follow-up studies, revealed that ingesting DDT residues thinned eggshells and harmed bird species. This, together with other studies, resulted in the US ban of the pesticide in 1972.[13]

The Freudian lure of eggshells, though, has still proven irre-sistible for some British men. In 1975, Mervyn Shorthouse showed up at the Natural History Museum at Tring, just outside of London, which holds the world's largest egg collection at over one million specimens. He'd been injured in an electrical acci-dent on the job, he said—though later "on the job" turned out to mean "hacksawing through a live cable whilst trying to steal it." He arrived in his wheelchair, making a compelling plea to the egg men on duty: exploring the egg collection, he said, would bring him such joy. Over the next several years he visited the col-lection eighty-five times, until a curator noticed him slipping cal-cified shells into his pocket. The curator called the police, who searched his car and found 540 eggs. The search of his house turned up 10,000 more. He had apparently been selling eggs to other collectors.[14]

That situation is not usual, as Mark Thomas—head of investigations for the Royal Society for the Protection of Birds (RSPB)—told me. Most people who steal eggs don't rob muse-ums; they collect them like the Victorian oologists did, from nature. Only now, collecting eggs from the wild is illegal. Short-house seemed motivated by money, but for most egg thieves, it's not about money or trading or even growing their collection. In modern Britain, the secretive and illegal hobby of egging is driven by obsession and by the rush it brings. A clutch of rare eagle eggs is a kind of Proust's madeleine: the sight and possession of it evokes the day the collector took it, the tree he climbed along with the physical danger of climbing it, the rush of discovery, and the steps he took to avoid the authorities.[15]

Egg collection may have lingered longer in the United Kingdom because it passed laws banning the practice comparatively late. The United States banned the wild bird trade in 1918, but Britain didn't follow suit until 1954, when the Protection of Birds Act made taking wild bird eggs illegal. Unless one can prove the eggs were collected prior to 1954, possession is a crime. What had been a schoolboy gateway to nature appreciation was now illegal, and it's no surprise that it took a generation or two to get the message.

In the 1990s, Operation Easter began. The slyly named sting operation brought together the police's newly created National Wildlife Crime Unit and the civilian conservation group, the RSPB. Their pursuit: illegal egg collection, a crime perpetrated solely by men. No one working in law enforcement or museums has heard of a woman with an egg collection, perhaps because those of us with ovaries carry around our collections internally. Stealing wild eggs is also a uniquely British crime. As Mark Thomas told me, even when the group hears of egg crimes perpetrated overseas, "It's always British people going on an expedition, really, mirroring two, three, four hundred years ago, going abroad taking eggs where nobody will really know what they are doing." In other words, these guys are cosplaying Victorian gentlemen in a post-Victorian world.

These men share a common profile. They are young to middle-aged men limber enough to climb trees. "But the commonality is spending time in nature when you were young at school," Thomas told me. Modern collectors hail from all walks of life, including dentists, lawyers, and a police officer who was later prosecuted. Between 2000 and 2020, the number of active collectors has dwindled, Thomas said, but the ones still clinging to the hobby are quite hard-core and secretive, since a prosecution could mean not just jail time but the loss of a lifetime's collection. The remaining collectors aren't tweedy professor types but young men "who look more like a football supporter than anything," Thomas said. One group of collectors in the industrial city of Coventry, he said, were "naughty kids at school, not top formers, and they would spend their evenings in the wood taking eggs, and it became competitive to get eggs." When they got cars, they began going farther afield for finds, which put them at greater risk of contact with authorities. Thomas can't say for sure, of course, but speculated that spending time in nature as kids might have been an escape from other circumstances. As middle-aged collector Michael Stockton put it in *Poached*, a documentary about egg theft, "You love them [eggs] that much. You covet them," and "I might stay out of trouble when I'm going out . . . doing stuff in the forest and woods and stuff. . . . The mountains and that not got police. . . . I feel safe and OK in the mountains." Obviously, the mountains do have some police, and that's why he was fined £2,000 in 2003 for disturbing nests and possessing egging equipment, but broader point taken: he didn't feel surveilled in the mountains. He felt free.[16]

Collectors consider themselves sophisticated rebels, performing the jewel heist of wildlife crime, a cat burglary of nature's next generation. In addition to all the usual egg-collecting skills, eggers must avoid the police, so they often blow eggs on the spot and bury them in tins in the woods for later retrieval during a less suspicious (non-spring) time of year.

Police stings sometimes verged on the spectacular. One target was the Jourdain Society, a group started by gentlemen collectors Walter Rothschild and Reverend Francis Jourdain in 1922. The society had a supper club where oologists gathered—and continued to gather even after the law barred wild-egg collection. In July 1994, authorities sent an undercover policewoman to one of their dinners as a honeytrap. Men trying to impress her showed off photos of rare nests. The police busted the gathering and used intel from the raid in the following months when authorities seized 11,000 eggs in seven counties, resulting in convictions and fines for many members.[17]

Sometimes the arrests felt pathetic. Police nicknamed one pair the "Abbott and Costello" of egg theft. Arrested in Scotland in 1997, they liked to film themselves robbing nests. One video showed a fulmar vomiting on them (a natural defensive behavior) in a doomed effort to guard her nest. The poor bird, literally sick over the theft of her future babies.[18]

During the 1990s, the RSPB accepted call-in tips from friends, relatives, the fed-up wives and girlfriends of offenders, and folks who overheard egg talk in restaurants. But the problem of egg collecting remained, and the rarer the eggs, the greater the adrenaline rush of taking them. Over the years, collectors had nearly wiped out several species, including the Slavonian grebe and red-backed shrike. In 2022 the Slavonian grebe had just thirty breeding pairs remaining in the United Kingdom, while the red-backed shrike had only up to three breeding pairs left in the region. Both species have "red" UK conservation status, indicating that they are urgently threatened, locally and globally. Unfortunately, mere fines didn't deter collectors; they would often pay and then reoffend.[19]

Until 2001, the RSPB's crime unit spent about half its time helping catch egg collectors, the most pressing threat to Britain's wild birds. After much lobbying by the RSPB, in 2001, the law changed to permit jail sentences for offenders. As Thomas put it, after the first arrests, "because it's such a tight knit network, the word immediately went around." Jail time worked well as a deterrent for many collectors but spurred the truly obsessed to go deeper underground. Since then, Thomas has found egg collections in a lot of weird places: secret compartments hidden in door frames; under a wood floor in a conservatory; in cabinets tucked away in attics; beneath gardening sheds; under beds and sofas; and stashed in a thermal pitcher meant for coffee that had been hollowed out and lined with padding. One repeat offender, Daniel Lingham, was a woodworker and had every surface in his mobile home on hinges that opened to reveal his collections, said Thomas. According to Julian Rubinstein's excellent *New Yorker* article on Operation Easter, when the authorities raided Lingham, he said, "Thank God you've come. I can't stop," indicating that egg thievery can move into the realm of addiction. The defense attorney on one of Lingham's cases told the BBC, "His issues started as a young boy. . . . The only way his father would give him praise was if he collected rare eggs." In 2005, Lingham received a twelve-week jail sentence after authorities seized his collection of over 3,600 eggs; in 2018 a tipster called the police after watching him in action. Authorities arrested Lingham in head-to-toe camouflage, with climbing spikes, a catapult, and 9 linnet eggs on him. "I've been a silly man, haven't I?" he asked the officers. Officials found more than 5,000 eggs at his house this time. Tragically, the collection included 681 eggs forming 164 clutches from rare birds. As the RSPB bulletin on the most recent arrest put it, he "had also collected 109 nightjar eggs by systematically plundering almost all the known Norfolk breeding sites."[20]

The Convention on International Trade in Endangered Species of Wild Fauna and Flora (CITES) treaty lists some five thousand species of animals as protected, sorting them into schedules in order of endangerment similar to the way governments use schedules to distinguish street drugs. In the United Kingdom, for example, LSD and MDMA are schedule-one drugs. According to CITES, peregrine falcons are schedule-one birds—and to collectors, these rare birds are catnip. Comparing heroin to, say, osprey eggs may not be that far off for certain people.

Another serial collector, Matthew Gonshaw, carries the distinction of being the only person banned from Scotland during the breeding season—for life. That's because he was an egg addict. First arrested and fined in 2001 while raiding a golden eagle's nest in the Outer Hebrides, he went on to reoffend and spent time in jail for egging in 2004 and 2005. But he couldn't stop. In 2011 authorities caught him *in flagrante delicto* on the Scottish Isle of Rum with eggs and a blowpipe on his person, tools that contained DNA evidence from several wild birds. The ensuing raid of his apartment revealed seven hundred eggs in hidden compartments and cabinets, including eggs of schedule-one birds like osprey and golden eagles. The poor golden eagles—their eggs had large holes in their sides, indicating they had been near hatching when Gonshaw lethally evicted them. Gonshaw, a clean-shaven, fit, handsome guy living on public assistance in London, meticulously budgeted for his adventures, down to the discount bus tickets to forests and packets of instant custard he ate while adventuring. As Thomas told me, the key with deterring Gonshaw from further egging was "figuring out exactly what made him tick. . . . With him, it was all about the freedom of being able to go to Scotland and roam around. . . . He didn't have a great deal going for him in his life and when he was in Scotland he was focused. . . . The buzz of going out and doing that and seeking that freedom was far greater than any punishment." So authorities destroyed his collection, seized his birding gear, and barred him from Scotland during breeding season.[21]

For Gonshaw, the loss of his gear, the loss of egging in his life, seemed devastating. In *Poached,* director Timothy Wheeler shows Gonshaw in a sparse apartment. He was clearly trying to move on with his life, with a newfound devotion to yoga. Like the other collectors profiled, he got . . . strange . . . when he talked about egg collecting. He presented himself as the reformed, yoga-loving ex-con, but at other times, he talked defiantly to the filmmakers. For example, despite his lifetime ban from Scotland, he said, "If I wanted to go up there, what could they do? If I walked into a police station, what could they do? Arrest me? Take me to court? If I wanted to I could still take all the eggs I wanted to in England, couldn't I? There's plenty of good birds that nest in England." Wheeler walked by a nature preserve with him, where he rattled off lists of all the species he once collected there. When I asked Thomas about Gonshaw's collection, he looked serious and sad. "I remember raiding him," he said. "I went through all his diaries." Along with hundreds of pages of flowery text about birds, Thomas said that one page simply read "need to get a girlfriend." Thomas added, "It's a very sad hobby."[22]

I find these men—collectors old and modern—fascinating as a metaphor. They rob birds, specifically female birds, of their vital resources, the things they love so much they will squawk and vomit to protect them. The collectors are killing potential new generations of birds, all so they can hang on to the ossified shells. Arguably, this is the story of colonialism, a quest to control the natural world by ordering and exploiting it. That the modern men suffering from addiction to wild-egg theft were almost universally British and predominantly working class, well, to me speaks to a crisis of class identity and masculinity. Wearing my unlicensed armchair psychology hat, I'd say it gestures toward nostalgia for the era of British dominance, a time when a British man could rest firm in the knowledge that the sun never set on his empire. And yet, in modern times, collecting the objects that rich British gentlemen once did is no longer acceptable. In fact, it's punished with jail time. Gonshaw's rather pitiable diary lament "need to get a girlfriend" cuts me to the quick. As Thomas pointed out, if you remove the egg aspect from Gonshaw's self-presentation, "if you remove that and just look at him, he's just a troubled individual."[23]

While collectors still find ways to steal eggs, the heyday of illegal egging has largely passed. Custodial sentencing deterred many casual collectors; prosecutions have dwindled. The RSPB spent as much as half of its time dealing with eggers in the mid-2000s, but now "it's a minuscule time per year," Thomas said. The United Kingdom's changing customs around egg collection mean that a new generation of collectors has not arisen to replace the old one. By the time of writing this, most people considered taking birds' eggs to be wrong, which wasn't true in 1990 or even 2000.[24]

Although eggers were once the most pressing danger UK bird populations faced, there have always been others, Thomas tells me. A custodial jail sentence effectively reduced the problem of egging, which freed up the RSPB to tackle other problems, such as climate change and raptor persecution. In Great Britain, people kill protected birds of prey, often to safeguard the tradition of grouse hunting on managed game preserves. Basically, both humans and raptors like to hunt grouse, so humans kill raptors presumably to preserve game populations for hunting.[25]

The Victorian mentality isn't quite dead. A few underground egg obsessives still lavish attention on their ovaloid treasures. But at the end of a hundred years of conservation movements, only one collector remains standing: the museum, with its drawers full of bird skins and eggs.

EGG RUSH

Egg dishes are a flavor of home. We want them the way our fathers or mothers or grandparents made them. An egg conjures memories of learning to cook for many of us, since making eggs is perfect for teaching a kid. Even if one disdains a straight scramble, the egg is a key ingredient in many comfort foods, including pancakes and birthday cake. For all these reasons, eggs carry a certain nostalgia. They remind us of paradise lost—the childhood that is done, the beloved elder chef who is buried, the metabolism that once tolerated syrupy breakfast carbs. And nostalgia has power. Would you kill to return to a yolk-kissed moment when a caregiver served up love on a plate? Some men would. And, indirectly at least, some men did during the Egg War of the Farallon Islands.

The Egg War began unofficially in 1848 with the Gold Rush. San Francisco started the year with a mere thousand souls, but over the next twelve months the population rose to twenty-five thousand. The city experienced scarcities of women and of food, particularly protein. Scaling up farms to provide for the local population proved harder than it seemed. Nobody could get large groups of chickens to survive there, and the technical solutions to this problem were decades off. Without chickens, of course, there could be no eggs. And without eggs, there could be no cakes, morning scrambles, pancakes, puddings, or muffins. As Napoleon once put it, "An army marches on its stomach," and a rootin'-tootin' army of miners in the Wild West doubly so. As gold poured into the city, the prices for fresh eggs skyrocketed. Out in the field, a single chicken egg might sell for $3, while in the city that same egg fetched the still exorbitant price of $1. Even without accounting for inflation, $12 to $36 per dozen eggs is ridiculously expensive. If we account for inflation, the miners paid something astounding—more like $427 to $1,282 per dozen. This explains the origins of Hangtown Fry rather well. According to legend, a guy who had struck gold wandered into the El Dorado Hotel in the mining supply camp of Hangtown (so nicknamed for its penchant for stringing up criminals). He threw down a bag of gold and demanded the most expensive meal the chef could make—which turned out to be oysters and eggs. If someone could bring good fresh eggs to San Francisco Bay, he would more than make his fortune.[1]

By most accounts, the first people to strike it rich were "Doc" Robinson and his brother-in-law Orrin Dorman. Doc, a pharmacist from Maine, had figured out that the Farallon Islands, home to hundreds of thousands of screaming seabirds, might provide enough eggs to finance a new pharmacy. So Doc and Orrin hopped in a boat and set sail for the Farallones, about thirty miles outside of San Francisco Bay.

As a location, the Farallones are pretty cursed; they are the sort of place a third-grade boy would make up to impress and gross out his classmates. Although to call them "islands" is a bit grand—they are jagged rocks of various sizes that stick up above the water. Those rocks are a legendary site of shipwrecks. Since Sir Francis Drake set foot on the islands in 1579, mariners have referred to the group as the Devil's Teeth, for their appearance, the rough seas that surround them, and their tendency to chomp on ships. One of the smaller islands is known simply as "the pimple," a rock six meters tall and sixty-five meters wide, with a whitehead of bird droppings on it. The sea salt blasts the rocks, and white crystals crunch underfoot on some of them. Before the Europeans arrived, the Ohlone people called the Farallones part of the land of the dead, specifically the part meant for the bad, dead people.[2]

If these weren't enough of a deterrent, great white sharks infest the seas surrounding the islands, which is unusual since great whites tend to only travel solo or in pairs. But around the Farallon Islands, they gather in numbers up to 150 sharks. It probably has to do with the large population of pinnipeds that also populates the Farallones. Once home to fur seals and sea otters until Russian and Boston fur merchants decimated local populations in the early 1800s, the islands also played home to elephant and harbor seals, as well as several species of sea lions—a veritable buffet for the great whites. The kelp flies, of course, also came to lunch on the pinnipeds. And kelp flies number among nature's vilest creatures. I will let journalist Susan Casey explain. For unfathomable reasons, she spent weeks on a rickety yacht moored offshore of the largest island in the 2000s while reporting for her shark book titled after the archipelago:

> They were at their peak now, a carpeting plague, crawling up pants legs and down shirt fronts, overwhelming a person's every moment outside. And these flies were not the cleanest insects—their preferred habitat is the inside of a seal's anus. The anus flies spent their time in one of three ways: tormenting us, tormenting the poor seals who had to house them in such an inhospitable place, and copulating with abandon in giant fly gang-bangs. This morning I'd counted a vertical stack of thirteen flies.[3]

I personally draw the line at anus flies.

Into the tumultuous, shark-infested, kelp fly–ridden water, Doc and Orrin sailed. They landed on the largest rocky outcropping, which is less than a fifth of a mile square. There, they came face to face with the overwhelming fact of life on the island—its birds. Hundreds of thousands of seabirds—gulls, cormorants, auklets, puffins, petrels, and most important, the common murre. These shrieking, squawking birds, packed shoulder to shoulder among the cliffs, laid great volumes of eggs onto masses of weeds and likely comprised more than half a million nesting pairs. Importantly, the common murre outnumbered the other seabirds. A type of guillemot, the common murre dresses all in black, save for its white belly. Each year, female murres lay a single pear-shaped egg with a tough shell. The eggs have background colors in the greenish blue range, with darker brown-black pencil squiggles and dots atop. Roughly twice as large as a chicken egg, with a bright red yolk and a white that stays translucent when cooked, murre eggs made a fine substitute in baked goods. When not eaten absolutely fresh, though, they leave an aftertaste of old fish. Eat a thoroughly bad murre's egg, and rumor had it that you'd spend three months getting the flavor out of your mouth.

Doc and Orrin scrambled up those slippery, excrement-covered cliffs and filled their boat with eggs. On the harrowing journey home through rough seas, they lost almost half their booty. But when they arrived in San Francisco, their half boatful of eggs fetched a small fortune of $3,000 (something like $100K in 2020 money). Doc Robinson used his share of the profits to build a pharmacy and the Drama Museum, a theater where he delighted locals with his impressions of New Englanders. He went on to become a pillar of the nascent theater community. But the trip had so terrified him and Orrin that nothing could persuade them to return. Word of their profits, though, spread quickly. The egg rush had begun.

Within a week, eggers swarmed the Farallones seeking their fortune. One enterprising collection of six men promptly formed the Pacific Egg Company (also known as the Farallon Egg Company, or simply the Egg Company) and, in keeping with the land-grab ethos of the time and place, staked their claim on the largest island. They fought off their foes, erected some outbuildings, and soon established brutal methods for gathering eggs. First, they'd rampage through one section of the egg fields, breaking every egg in sight, which ensured the freshness of the next day's harvest. Their crews had specialized gear—rope-soled shoes, often with spikes driven into them, to help them gain purchase on slippery cliffs. The egg man's uniform also included climbing ropes and special vests made of flour sacks with a drawstring waist and holes cut for the head and arms. Eggers deposited their cargo into a deep slit in the vest's neck, which allowed them to carry up to eighteen dozen eggs without a basket. On the cliffs, keeping hands free was key. When fully laden, the eggers resembled lumpy Santas and would return to a collection point at the base of the cliffs. They would kneel deeply over a basket, almost as if praying, and let the thick-shelled eggs pour from their chests.[4]

The work required desperation or nerves of steel, probably both. The company employed up to twenty-five men at a time, often new immigrants with little to lose. The treacherous egging season ran from May to mid-July. A simple slip on the cliffs could send a worker into the shark-infested brine. And then there were the gulls. According to an 1874 *Harper's* article on the eggers, "These rapacious birds follow the egg-gatherers, hover over their heads. . . . The egger must be extremely quick or the gull will snatch the prize [the egg] from under his nose. So greedy and eager are the gulls that they sometimes even wound the eggers, striking them with their beaks."[5] To avoid frequent scalp injuries, many of the eggers carried clubs, which they swung around their heads.

With the Pacific Egg Company in control of the largest island, local fishermen and other fortune-seekers ventured out to the smaller boulders. A local newspaper ran a story on one such pair that ended up stranded on a rock for six weeks in 1899. Stormy seas foiled at least three rescue attempts, leaving the men to survive on raw eggs and the meager supplies rescuers could land. Returned to safety, one of the emaciated castaways told the local newspaper, "I will never again be able to look at a murre egg without disgust. We have had several fights with the sea lions." A dramatic drawing of a man with a bat fending off a ferocious pinniped accompanies the story.[6]

Throughout the 1850s, the Farallones' yearly egg season brought armed struggles as the Pacific Egg Company vied for control of its veritable gold mine. It fought off gangs "armed to the teeth," according to an 1859 *Alta California* article. Battles raged on land and sea, as hijackers attacked boats ferrying eggs to the mainland. One group of rival eggers spent several days hiding in their boat inside Great Murre Cave beneath the largest island, where guano continuously rained down on them and the ammonia buildup killed several men. A government force sent into the fray in 1860 found themselves so outnumbered and outgunned that they "thought it prudent" to return home without engaging.[7]

If it wasn't rival gangs of eggers testing the egg company, it was the government. In 1855 the US government seized the land and built a lighthouse there. It refused to recognize the egg company's claim to the land but allowed it to keep up its rapacious methods so long as it didn't interfere with lighthouse business.

The federal government fixed pay for all lighthouse keepers at a paltry $450 to $600 per year. Not bad if you lived in the East, but in the inflation-happy epicenter of the Gold Rush, domestic servants could earn nearly that in a *month*, plus room and board, and all without having to live in a nightmare hellscape of guano and bird shrieks. Nerva N. Wines, the first lightkeeper, who served from 1855 to 1859, became a stockholder for the egg company and let them run amok so long as he received his dividend. His successor, Amos Clift, had a better scheme. Clift had taken the job for the express purpose of commandeering the eggs. As he wrote in a letter to his brother, "If I could have the privilege of this egg business for one season, it is all I would ask [and] the government might then kiss my foot." Clift boldly leased egg-gathering rights to various parties and remained the keeper through the season of 1860, when the Lighthouse Board fired him for corruption. Without Clift managing the many competing parties, the conflict heated up.[8]

It started off with the egg company posting signs barring the lightkeepers from certain parts of the island. Next, eggers busted up government roads, and an armed group captured four lightkeepers and tried to eject them from the island. Later that same year, someone assaulted an assistant lightkeeper. The situation had gotten more out of hand than usual, and the government struck back. The regional superintendent of lighthouses, Ira Rankin, had a pragmatic streak and realized that so long as the egg rights to the land were up for grabs, the assaults, stabbings, intra-egger battles, and graft would continue. So he decided to crown the original egg company ovary overlords of the Farallones and to hell with whether it was technically legal. (On paper, at least, the land belonged to the US government.) And Rankin would support the egg company using government power.[9]

A freelance egger named David Batchelder took powerful exception to this move and made repeated, armed attempts to take the island with increasingly large numbers of men, Italians, a detail the papers loved to include. During the start of the 1863 season, he and his men built a house and a stone fortification on the island. Rankin responded with an armed customs ship that laid siege to Batchelder's operation. The government removed four men, five shotguns, a rifle, and assorted other weapons. But Batchelder was not easily deterred. Two weeks later, he returned with at least thirty men, who captured several Pacific Egg Company employees. Again, the same customs ship arrived and landed three boats of men to round up the egg rebels, plus their arsenal of twenty-one firearms. At some point, Rankin realized that a few lightkeepers must be in league with Batchelder, so he sent a sternly worded letter threatening to fire anyone assisting the upstarts. Rankin also ordered the customs ship to patrol nearby waters, questioning any boat headed to the Farallones and boarding it if necessary.

Batchelder was rumored to be gathering forces for another try. On June 3, 1863, three sloops dropped anchor off the coast of the main island. They contained Batchelder, twenty-seven armed men, and a four-pound cannon. Isaac Harrington, the egg company foreman, met the boats at the landing, a wooden derrick built over the inhospitable shoreline. He howled across the waves that the rebels would land "at their peril," and Batchelder yelled that they would come ashore "in spite of hell." Everyone spent a tense night, the egg company men camped on the landing and Batchelder's men carousing in their boats. At daybreak, the rebels sent one boat in for a landing, and everyone opened fire. When the gunshots and feathers settled twenty minutes later, one of the company men was dead with a hole blasted through his stomach, and Batchelder's men beat a hasty retreat, leaving a sloop behind. Five of Batchelder's men had injuries, including one shot through the throat, who died at a hospital a few days later.[10]

After Batchelder's grand defeat, the rivalries among egg gangs died down, though tensions between the Pacific Egg Company and lightkeepers remained high for several more decades until an 1881 executive order barred commercial collection on the islands. Twenty-one soldiers arrived to evict the egg company from the land permanently. Informal egging and selling by the lightkeepers continued till the end of the century but eventually ceased as the rising supply of chicken eggs made it far less profitable.[11]

The Farallon egg trade lasted for a half century, with tragic ecological consequences for the birds. Estimates vary from source to source, but at the beginning of the egg rush, the company likely shipped around 900,000 eggs per year. Fifty years later, that number was closer to 150,000 eggs shipped, a sixth as many. The unchecked smashing and stealing of murre eggs had a predictable effect, decreasing the murre population by about 95 percent, from a high of 400,000 to 600,000 before egg gathering to a lean 20,000 birds at the trade's conclusion. Later environmental degradation—multiple oil spills, shipping lanes, falling numbers of tasty sardines, to say nothing of an underwater nuclear waste dump—further diminished the number of murres to a mere 6,000 by the 1950s. Since then, thanks to conservation efforts, numbers have greatly recovered, hitting 100,000 in 2000 and 250,000 to 300,000 in 2020.[12]

Humans have done far worse in pursuit of eggs. Before Doc Robinson retrieved the first boatload in the Farallones, before the Gold Rush entirely, the great auk flourished. A large, docile, penguin-looking bird, the auk was a member of the common murre's biological family. Like murres, great auks congregated in large groups—a move that would turn out to be foolish—to lay their eggs on the bare rock of sea islands. Black with white bellies like the murre, this oversized, flightless cousin bred on rocks off the coast of Greenland, Newfoundland, Iceland, Massachusetts, and Scotland. They had the minor misfortune of moving and breeding slowly—females laid only one egg a year— and the greater misfortune of soft feathers, tasty flesh, and fat that made a fine fuel oil. Laws going back to the 1550s tried to protect the birds, but they proved too easy to catch and too useful. One of the sadder passages I've read is the 1794 description of an auk hunt by Aaron Thomas of the HMS *Boston*:

> If you come for their Feathers you do not give yourself the trouble of killing them, but lay hold of one and pluck the best of the Feathers. You then turn the poor Penguin adrift, with his skin half naked and torn off, to perish at his leasure. This is not a very humane method but it is the common practize. While you abide on this island you are in the constant practice of horrid cruelties for you not only skin them Alive, but you burn them Alive also to cook their Bodies with. You take a kettle with you into which you put a Penguin or two, you kindle a fire under it, and this fire is absolutely made of the unfortunate Penguins themselves. Their bodies being oily soon produce a Flame; there is no wood on the island.[13]

The human cruelty blows me over—to skin animals alive or boil them alive on a fire made of their fellow birds. It's a vision of hell.

By 1800, many of the great auk habitats had been destroyed. The scarcity of great auks made their eggs more valuable to oologists, who sent collectors out to snatch eggs and skins, which further decimated their populations. In 1844, three Icelandic hunters visited Geirfuglasker, a coastal island, to secure some specimens for a merchant. On June 3, they found the last known nesting pair of great auks, strangled them, and deliberately smashed the last egg with a boot. One of the hunters later described the scene to a researcher: "I took him by the neck and he flapped his wings. He made no cry. I strangled him."[14]

Early in 2020 I visited a collection of great auk eggs at the Museum of Comparative Zoology at Harvard University. Collection manager Jeremiah Trimble, red-headed and with a hint of stubble on his cheeks, had an obvious love of the natural world; he lit up when I asked him about birding. He ushered me into the large windowless room that hosts the museum's enormous collection of nearly 500,000 specimens of eggs and skins. Taxidermied birds perched atop the grayish cabinets that packed the room. I had a list of eggs I wanted to see—the glossy iridescent green-blue of the South American tinamou inside a clear plastic box within its drawer; the Tic Tac–sized hummingbird eggs, displayed inside nests that were, as Trimble explained, held together with cobwebs, like some fairy house. I saw globed white owl eggs, and the beautifully pointy and speckled egg of the common murre, including one collected in the Farallones. The egg of the elephant bird of Madagascar had recently been on display and lived in a cabinet. Such birds once laid the world's largest eggs—the size of watermelons. Though historians now debate the cause of their extinction, which occurred sometime between 1000 and 1200 CE, one theory suggests that humans simply ate them to death.

After fitting the enormous beige-brown balloon into its box, Trimble showed me a very special cabinet on the other side of the room: a drawer of extinct species. He showed me the oblong golf ball of what was the United States' only native parrot species, the Carolina parakeet. Down the row from it lay a small, glossy white egg from the passenger pigeon, a species too tasty for its own good. Perhaps the world's largest collection of great auk eggs lay nearby, more than a dozen of them, the supersized version of the common murre eggs.

Though I try to be open-eyed about the depravity of the era of exploration, I confess I do feel nostalgic for it sometimes. Not for the bad medicine or the narrow range of roles women and people of color were allowed to play in public life, but for the global sense of adventure and for the natural world that has been lost. How wild that the United States once had an exotic parrot species, with its orange head and teal body. I wonder about the people back then: did they count themselves lucky to walk the Earth at the same time as the Carolina parakeet or the great auk? Or did they simply see birds as commodities for exploitation or as mere scenery?

Peeking inside the drawer sparked a mix of emotions: illicit delight in seeing something rare mixed with solemnity since all that survives of these species are these remains locked in basement drawers. Of course the Victorians did not know how lucky they were to walk the Earth with the Carolina parakeet any more than I knew, when my father-in-law came over to play with my baby one morning, laid him on the carpet floor, and said, "Coochie coochie coo—you're not going to see any white rhinos, are you? No, you're not" to my child on the day the last male white rhino died. The white rhinos are lost. There are no more elephant birds. There are no more night herons or dodoes or Reunion ibises. I thought of that sepulchral drawer again nearly a year after my pilgrimage, when the US Fish and Wildlife Service announced the extinction of twenty-three more plants and animals, including Bachman's warbler, a symphony in yellow, black, and moss-green. We have made the same mistakes over again, and as I remember the large, old great auk shells in the cabinet, I think perhaps the birds are not the foolish ones after all.

EGG MONEY

When I was nineteen, my parents put me on a plane to the West African nation of Togo to visit my cousin Mandy. I had no idea what to expect beyond red dirt and the advice from my aunt, who had visited her daughter over Christmas, that it would be "a little bit like camping." I flew into Ouagadougou, in Burkina Faso, where my cousin picked me up, and we bused south, across the border to Togo. Whenever the bus stopped, women gathered by its windows, selling plastic baggies of water and juice, mangos, and hard-boiled guinea fowl eggs.

Finally, we arrived at the small village Mandy had called home for the last few years and settled into her two-room house, built per Peace Corps regulation on a cement slab. I met her dog, named for the local beer, and her cat, named for the local cheese. The night after I arrived, my cousin asked her best friend in town, Amina, to make dinner for us. Amina raised guinea fowl—black birds with white polka dots, blue heads, and red combs—and Mandy had bought one from her to cook for the meal. We walked down the clay road through town, past the lone, generator-electrified house, to the compound Amina shared with her husband and his other wives. Mandy had already pointed out that polygamy did have some advantages. For example, one of Mandy's neighbors had twin toddlers, and it was incredibly convenient to be able to hand one of them to your co-wife when they both started crying.[1]

Amina had made us a stew of guinea fowl, spicy tomatoes, and beans over rice. We ate sitting on blankets on the ground, a shared bowl for my cousin and me, one for Amina, and other shared bowls for various members of the household. In contrast to the United States' herculean meat portions, here Mandy and I had a single wing to split. You are supposed to eat it last and savor it, kind of like a protein dessert, Mandy said. Although she had coached me before the meal on how to eat with my fingers, I struggled to strip the meat off the bones one-handed and gladly accepted some advice and help.

I could see why my cousin liked Amina. She was a lively woman, full of laughter. Taller than either of us, Amina had the marks of her ethnic group over each high cheekbone. After dinner, she brought out a treat: several large glass bottles of Fanta and Coke. Mandy translated from French for me as we talked and laughed under the stars. In the velvety darkness we could see the Milky Way and constellations like my star sign, Scorpius, not visible in my urban hometown. When we walked past the house with the generator on the way home, the light was bright enough to hurt my eyes.

Little by little over the next few days I learned the true story of the dinner. Mandy had bought one of Amina's guinea fowls and commissioned her to kill it and cook it for us. In addition to feeding us all, it put money in Amina's pocket and protein into the stewpot we shared with her family. My visit coincided with the start of a drought and accompanying food crisis; the evidence surrounded us. Many children had bulges around their navels from a lack of protein in their diet. And, despite owning guinea fowl, Amina's family didn't often eat them. They were too valuable at market to squander at home.

Amina's flock had provided for her over and over again. A women's rights group had given her the capital to build the coop. She raised the hens cheaply in her country village and took them to the city to sell for a higher price. This side job augmented the money she also earned as a seamstress. She must have been in her early thirties when I met her, but she had already given birth to six children. And on one trip to town to market her birds, she did something bold. She used money—her own money—to pay for a five-year birth control implant in her arm without her husband's permission. When she returned home, she let him know. Take it out, he said. I can't, it's under my skin, she said, and there is no doctor here who can safely pry it loose.

Closer to the time of my visit, Amina made the trip to town to sell the birds during school week, leaving her eldest daughter to take care of the other kids. How could she have left the kids, her duty, during school week? Her husband was furious. He hit her and threw her out of the compound, but he couldn't make her homeless. She could have stayed with friends and family but decided instead to shame him. The compound belonged to him, but the coop belonged to her. She had her daughters clean it so she could sleep there. Her skill at breeding and rearing guinea fowl gave her poultry money, and that money offered her options she would not otherwise have had.

◉

Poultry-rearing has a long history as a vehicle for empowerment. For most of that time, chickens earned only modest profits. Flocks topped out at around a few hundred birds: any more and the barnyard's brutal pecking order resulted in too many deaths. A flock's rate of increase also had limits, since only so many eggs fit under a bird's rear. Chicken husbandry relied on a strange quality of domesticated hens, which lay one egg every day or two but don't begin to incubate them until a pile accumulates. If you remove the eggs as they are laid, however, the hens will produce up to three hundred eggs per year until their egg productions slows so much that they are considered "spent."

In the 1600s, the upper classes in the United Kingdom and American colonies enjoyed beef and lamb, while eggs were a "white meat" of the poor, along with butter, milk, cheese, and sometimes chicken. The birds didn't keep the best company: the scrawny chickens rifled through the waste of other livestock for food and acquired a reputation to match. The colonists called them "dunghill fowl" and, later, "yard birds." Likewise, in colonial animal husbandry, beef and pork were the real cash makers; chicken was an afterthought, a garbage bird suitable for humans at the bottom of the pecking order.[2]

In the antebellum South, people who were enslaved capitalized on this oversight in the little time they called their own. As culinary historian and Queens College professor emerita Dr. Jessica Harris explains in her book *High on the Hog: A Culinary Journey from Africa to America*, "Some saved and foraged seeds and tended gardens by moonlight or fished and hunted nocturnal animals like possum; others raised yard fowl for their eggs or hogs for their meat."[3] These activities meant both food and side money. The 1741 slave code of the Carolinas, widely adopted throughout the South, went on to forbid enslaved people from owning hogs, cattle, or horses—expensive animals that could quickly generate revenue. However, as Emelyn Rude writes in her fowl history *Tastes Like Chicken*, "Conveniently missing in the text of the law was the chicken; dunghill fowls were of so little value to colonial landowners that they didn't record them in their inventories after all. (Such an allowance seems to be a universal commandment of slave codes; in ancient Palestine rab-

bis forbade low-level workers to sell wool or milk but chicken
and eggs were just fine.)" Chickens produced multiple commodities, including feathers, meat, and eggs. Sold to locals or even
to slaveholders, these products could bring in some money for
needed goods or to save in hopes of buying one's freedom.[4]

This historical fact—that enslaved people could keep chickens but not other sorts of animals across much of the South—
is the root of a racist stereotype mocking Black Americans for
a love of chicken. The iconic southern-fried chicken is a cross
between Scottish fried chicken, which came over with the landed
gentry, and the seasoned fried chicken of West Africa, adjusted
for English and Scottish palates. Black women—enslaved and
free—have often been in charge of cooking these labor-inten-
sive dishes. Before processing plants functioned as prep kitch-
ens, fried chicken required a lot of work. You needed to catch
the chicken, kill it, bleed it, scald it, pluck it, gut it, and butcher
it. Only then did you prepare to cook it for a specific dish—a lot
of work for only a couple of pounds of meat.

In *Building Houses out of Chicken Legs: Black Women, Food,
and Power*, Dr. Psyche Williams-Forson tells how Black women
made Gordonsville, Virginia, the "Fried Chicken Capital of the
World" in the late 1800s. These enterprising women took advantage of the town's two railroad lines by selling fried chicken to
customers through the train-car windows. In an era before dining cars and refrigeration, the dish traveled rather well. Black
people with few freedoms ingeniously took a devalued food,
used it to support themselves, and turned it into a touchstone of
American cuisine.[5]

In Depression-era Minnesota, my grandmother also took
care of chickens on the farm where she grew up. She regaled
us with stories about attempts to out-weasel the weasels, which
would sneak under the henhouse, reach up through the cracks

between the floorboards, yank down the baby chickens, and then eat whatever dangled. She was left to clean up the mess in the morning.

Farm women like her turned eggs and chickens into food for the family or into money when they sold them. For most of US history, married women had few property rights. That situation began to change in the mid-1800s thanks to legislation, but the process was slow. (Even as late as 1970, many married women couldn't get credit cards without their husband's approval.) Farm husbands tended to control the finances, but wives tended to control household budgets. Women had "pin money," either a husband-given allowance for niceties like hair pins or jewelry or leftovers from the household budget after the bills were paid. Pin money was a gift; egg money was what a woman earned on her own.

Through the late 1880s, chickens remained the domain of women and children. After all, the birds ate table scraps and stuck close to the house. As money makers, chickens and eggs didn't threaten the masculinity of the husbands (at least at first). Men did the "real farming" of big farm products like beef, pork, and wheat. Yes, it was OK for the little lady to have her hen hobby and make a few dollars on the side, building up her "nest egg." The concept takes its name from an actual object called a nest egg: a wooden decoy placed into nesting boxes to encourage hens to start laying. Egg money financed extras, like Christmas presents, music lessons, or over many years college savings for a child. Sometimes it formed an important slice of the family's budget and paid for household expenses like clothes or food. In many places, women bartered eggs and meat for groceries and other goods. Egg money could also serve as an emergency fund in case of bad crop yields or other financial disaster.[6]

As women and people of color began testing and making

different and better feeds, chickens became larger and brought in bigger sums. Farm journalist Laura Ingalls Wilder, later of *Little House on the Prairie* fame, reported extensively on the chicken-and-egg trade. In one article, she writes that the women of Mansfield, Missouri, shipped $58,000 worth of eggs in 1916, about $1.5 million in 2021 dollars. Not bad for a little farm town of fewer than a thousand people. As Wilder wrote, "I wonder if Missouri farm women realize the value in dollars and cents of the work they do from day to day in raising farm products for the market?" A single local woman had made $395 off her hens (worth about $10,000 in 2021), Wilder wrote, more than double the woman's annual household expenses. The little lady's "hen hobby" sometimes out-earned her husband's "real" work.[7]

Raising chickens remained the work of women, children, and people of color for as long as their profits constituted extra cash—about through the 1930s. But as the United States industrialized, the science of poultry became more efficient and profitable. With larger sums at stake, white men began to take over the chicken biz.

This movement from small farms to large commercial ones, along with the accompanying rise in profits, required getting around natural limitations. The size of a chicken's butt presented one such hurdle, as it determined how many eggs could be hatched at a time. Hens may lay plenty of eggs, but that only helps increase a flock quickly if you can find a way around the thorny problem of artificial incubation. Eggs are sensitive little things that require an exact temperature (97°F to 101°F) and a specific level of humidity (50 to 55 percent). Hens achieve this by tearing feathers off their breasts—the origin of the saying "feather your nest"—to create a bare swath of skin called the brood patch. The bare skin increases contact between a chicken's body and her eggs, which keeps them warm and moist. In addition, hens turn their eggs each day to help keep the yolk centered. Otherwise, embryos may fuse to the inner shell membrane.

Before electricity, thermometers, or humidifiers, meeting an egg's exact requirements proved tricky. But the ingenious ancient Egyptians managed to create a hatchery so productive that European visitors viewed them as semimystical egg factories, miraculously yielding chickens from ovens and piles of dung. The Egyptians created their mythical ovens around 323 BCE, about the same time chicken became a dietary staple there. (The Chinese followed close behind, inventing an egg-hatching method by 246 BCE, but they mostly used it on duck eggs.) The Egyptian hatchery consisted of a two-story conical oven. The upper level contained a smoldering, humid dung fire that gently heated thousands of eggs on the floor below. Attendants rotated the eggs daily and monitored the fire. They pressed eggs to the sensitive skin of their eyelids to check for temperature. Workers in rural Egypt used these same methods to hatch eggs as late as 2019. The clever incubator ovens remained a closely held national secret for thousands of years. In 1750, French traveler and entomologist René-Antoine Ferchault de Réaumur toured the egg incubators and brought the technology home to France, where it quickly flopped. France's cooler climate demanded hotter ovens, and the increased fuel use wasn't cost-effective. The truly efficient modern incubator was still more than a century away, but it would revolutionize the poultry industry.[8]

In 1878 a Canadian inventor named Lyman Byce moved from Ontario to Petaluma, California, a place up the river from egg-starved San Francisco. In Petaluma, he met dentist Dr. Isaac Dias, who was playing around with an egg-incubator prototype, although he failed to keep temperatures sufficiently steady. Byce solved the problem for him with a coal oil lamp and a mechanical regulator. Dr. Dias filed for a patent but died in a hunting acci-

dent before it could be granted. That left Byce with sole ownership of their world-changing invention, and he began producing
the incubators en masse. The Byce incubator hatched an incredible 90 percent of eggs successfully—a thirty-point improvement
over the best East Coast machines. Each foolproof incubator
held 460 to 650 eggs and only required human hands to turn
them each day. Christopher Nissom, another Petaluma farmer,
bought a Byce incubator and opened a hatchery. He sold chicks
to nearby farms and devised a way to send them live through
the mail.[9]

Thanks to the invention of incubator and hatchery, the San
Francisco region, which once longed for eggs, suddenly became a
world center of poultry. In 1880 Petaluma shipped 95,000 dozen
eggs to the city. Thanks to the Byce incubator, by 1915 they were
shipping 10 million dozen all over the world and had become,
according to food historian Dan Strehl, "the undisputed world
leader of the chicken and egg industries." Petaluma's mayor
declared it the "Egg Basket of the World" and launched National
Egg Day on August 13, 1918. American camp ensued. An Egg
Queen presided over a chicken parade, bird rodeo, horse race,
and more. My favorite bit of extravagance, though, is a bizarre
1930s promotional video from the event that shows a cadre of
attractive white women doing calisthenics in an oversized frying pan. Later, they help a chef make a giant omelet in the pan,
smiling while he gives directions and they stir. To the event's
deep detriment, it also included humor spiked with racism. The
town inducted members from the San Francisco Ad Club into
the "Order of Cluck Clucks Clan." They donned rooster caps
and crowed.[10]

Petaluma's fortunes peaked in the mid-1940s when it shipped
more than 42 million dozen eggs. But the advent of battery
cages—stacked wire cages stuffed with chickens—changed Pet-

aluma's fortunes again. Battery cages increased yields by packing more chickens into smaller spaces. The cages also got the birds off the ground and let their excrement drop down, translating into cleaner eggs for consumers. But for Petaluma's many small chicken farms, the cost of upgrading simply proved too much. Within a few decades, the industry faded from the area, and the center of egg production moved to enormous sheds elsewhere.[11]

Meanwhile, in Delaware, farm wife Celia Steele accidentally invented the broiler industry in the 1920s after a mailing mistake brought her five hundred live chicks instead of fifty. She and her husband raised them in a single coal-fired room, and to her surprise, it worked. The next year she ordered a thousand chicks on purpose, and ten thousand the year after that. In 1926 her husband quit his job, and her business became theirs. Word spread quickly among the neighbors, and the region became—and remains—a center of the US poultry industry. African Americans fleeing the South during the Great Migration provided cheap labor that fueled the new business model, but racism prevented Black people from benefiting from the new industry beyond employment. As Emelyn Rude notes, "By the 1930s, blacks comprised nearly thirty percent of the Peninsula's population, but almost none of them owned growing facilities themselves. Overt racism among the Peninsula's god-fearing Methodists prevented African Americans from owning farmland."[12]

As the chicken industry grew ever more profitable, white men became a greater force in it. A look at the executive profiles of modern chicken conglomerates like Cal-Maine Foods, which produced 19 percent of all US chicken eggs in 2021, reveals a leadership team that is overwhelmingly white and male.[13]

As for the birds themselves, well, the legal loophole in the slave code that allowed enslaved people to own chickens had other far-reaching consequences. Enslaved people couldn't own livestock, and so chickens were legally not considered livestock. The Code of Federal Regulations specifies that chickens and other birds belong to an entirely different category: poultry. That means essentially that animal cruelty laws that apply to livestock do not apply to chickens. According to Veronica Hirsch's overview of chicken law for Michigan State University's law school, food animals are lightly regulated in general, and "all the major farm animal protection laws exclude chickens. . . . From an animal welfare perspective, there are no federal regulations regarding the breeding, rearing, sale, transportation, or slaughter of chickens." In other words, the books I read to my child about farms, which always feature happy, free-range chickens tended by schoolchildren, are mostly lies.[14]

They are lies I am grateful for. As an adult, I am barely prepared to think about gassing male chicks on the day they hatch, or conveyor belts feeding them, alive, into a woodchipper. The female chickens are consigned to lives of reproductive servitude, confined in cages where they can't perform normal chicken behaviors: according to Hirsch, they can't "walk, fly, perch, preen, nest, peck, dust bathe, or scratch for food. . . . Hens may not even be able to stand up and their feet may grow into the wire floor of their cages." Animal behaviorist Temple Grandin writes that hens in battery cages face quality-of-life issues like "forced molting by food restriction," which basically reboots a hen so she can lay another three hundred eggs, as well as "osteoporosis, which causes bone fractures." It's hard on a hen's body to generate so much shelled calcium, and if the hen doesn't get it from her diet, it leaches out of her bones. To prevent chickens from pecking one another harmfully or fatally, farmers slice their beaks off with hot knives when they are only a few days old.[15]

My favorite greengrocer grows his own produce on two local farms. In the middle of the pandemic, I went to pick up a grocery order and was shocked to do so at a large henhouse. Until a few months prior, apparently, the store had also kept their own chickens and sold their eggs. I sat down with the farmer to ask him about chicken care. He told me that as a kid, chicken-tending and in particular egg-gathering were his disliked chores and that debeaking hurts the chickens about as much as trimming your fingernails. He also argued that free-range birds are mostly an expensive way to feed local raptor populations and drive up egg costs because the farmer must accept more losses. But I still can't help but wince at the idea of debeaking.

As far as I can tell, chicken biology doesn't bear out the comparison between birds' beaks and fingernails. Fingernails don't have nerves in them, but according to a 2020 paper by Australian agricultural professors published in the journal *Animal*, "The tip of the beak has a rich supply of nerves and sensory receptors," and the inside of the beak includes taste buds. Beaks help birds delouse themselves, preen, and, of course, enjoy their food. The paper goes on to call beak trimming "painful" and states that the conventional hot-blade trimming results in "acute and chronic pain" and is linked to reduced feeding. Trimming beaks with a newer technique that employs lasers apparently causes less pain and doesn't result in reduced feed intake. Beak trimming does do what it is advertised to do, though: it reduces feather pecking, aggression, and cannibalism in flocks, as birds get excited by the sight of blood. Much of Europe—the Nordic countries plus Austria, Germany, and Switzerland—has banned beak trimming.[16]

I'd like to believe that the eggs I eat come from happy chickens, but at the low price point folks in the United States enjoy, I confess I fear looking too closely. I suspect that if eggs come from big flocks, producers must make Faustian decisions about what's less cruel—debeaking the poor birds, which causes immediate and potentially chronic pain, or allowing them to keep their beaks at the price of chronic stress to the flock, poor feather condition, and, well, letting them peck each other to death.[17]

When I visited Mandy in Togo, before our dinner with Amina, she offered me the opportunity to kill the bird we would eat. It made her feel more connected to her food, she said, and it might be a rare opportunity for me to do so. She had done it only a few times, but if I didn't want to, Amina would do it. I imagined the glossy black feathers in front of me, my hands around its neck to wring it, and I passed on the opportunity. I regret sparing myself that moment of ethical reckoning. Me, on one side, an omnivorous mammal, the bird on the other—a rare opportunity to come face to face with the actual toll of my consumption, a chance I may not have again given my privilege in a country with a long, industrialized food chain. I do not have to personally witness the beak-trimmed birds sleeping atop one another in laying cages nor what happens inside an abattoir. But knowing that behind my omelet—somewhere blessedly offscreen—a tortured hen lingers has encouraged me to buy pastured eggs through a local meat-share. I pay a little more for the privilege, but it assuages my guilt.

As for small producers, they still flourish in West Africa. Billionaire philanthropists Bill Gates and Melinda French Gates have promoted chicken-rearing for people in the developing world. Chickens are cheap and easy to raise, represent an investment with a good return, and provide nutritious eggs to children. In a 2016 essay, Bill Gates states that the Gates Foundation's goal is "to eventually help 30 percent of the rural families in sub-Saharan Africa raise improved breeds of vaccinated chickens, up from just 5 percent now." Chicken-rearing is still women's work in many parts of the world, which means chickens put money in women's pockets. As Melinda French Gates wrote in a companion post, "I come across a lot of statistics in my line of work and maybe the one I've been most impressed by is this: When a woman controls the family's income, her children are 20 percent more likely to live past the age of 5." International development circles know chickens, she wrote, as " 'the ATM of the poor,' because they are easy to sell on short notice to cover day-to-day expenses"—expenses like birth control, self-determination, and a coop of one's own.[18]

5

EGG GURUS

The egg is a chef's ultimate test. According to legend, the hundred folds in a chef's toque represent mastery of a hundred egg dishes. When I ask culinary heartthrob Jacques Pépin about eggs, he quotes a nineteenth-century French cooking bible: "The egg is to cuisine what the article is to speech." That is, you can't do a thing in the kitchen without one. In this, the masters agree. Auguste Escoffier's legendary cookbook notes "there are very few culinary recipes that don't include eggs." Food science megastar Harold McGee writes that eggs' "contents are primal, the unstructured stuff of life. This is why they are protean, why the cook can use them to generate such a variety of structures." The word "protean," of course, comes from the Greek god Proteus, an old man of the sea, and means that something is as changeable and versatile as water. It is a good word for an ingredient that can clarify soups, foam meringue, raise baked goods, and so much more.[1]

There is just something about eggs. J. Kenji López-Alt earned my undying gratitude for his double-blind experiment hard-boiling seven hundred eggs for the *New York Times*. When I get the cookbook author and my personal culinary hero on the phone, he tells me, "Eggs have this mystique about them because they are so versatile and sensitive." That versatility translates to more egg dishes than colors in the Pantone catalog—and each person has their specific favorite shade. People have Opinions about eggs. Jonathan Swift knew it when he used them in his novel *Gulliver's Travels*. The nation of Lilliput had fractured—and gone to war with its neighbor Blefuscu—over whether to crack a boiled egg at the big or little end first. Kenji knows how deep egg preferences go too. "Eggs are always *the* most popular subject I write about," he said. "It's guaranteed that people are going to read it and talk about it and argue about it." Not only do people have preferences, but they are experts in their own right. "It's the first food that most people cook," he adds, and it's a food so commonly eaten that "people have more expertise in cooking eggs than in most things."[2]

I wrote one of my college essays on the egg chapter of my parents' 1974 edition of *Betty Crocker*, which began, "The man you marry will know the way he likes his eggs. And chances are he'll be fussy about them. So it behooves a good wife to know how to make an egg behave in six basic ways." The housewife's toque, presumably, only includes six folds.[3]

In addition to having Opinions and Feelings about the old version of *Betty Crocker*, I have definite ones about how to cook an egg. I enjoy many styles of eggs, but there is one thing I can't abide: rubbery, browned, hard eggs. (A crispy fried edge is the exception that proves the rule.) My basic philosophy is "pudding texture or get off my lawn." But the loose French scramble that is perfection to me is a Lovecraftian nightmare for my friend Jason, who prefers his eggs quite solid. To my vegan friends, meanwhile, eggs might as well be ashes: they taste only of suffering.

We owe our highly individualized, special snowflake preferences to the egg's versatility, which comes directly from its chemical makeup. I'm tempted to call an egg a holy trinity of ingredients—the white, the yolk, and the two intertwined. The white and yolk, protein and fat, are the yin and yang of culinary building blocks. In fact, traditional Chinese medicine literally considers them so. To explore an egg's versatility, let's analyze the structure of each part.[4]

Let's start with the yolk, essentially a balloon filled with water and fatty spheres, the spheres packed so tightly that they flatten into disks. These microscopic disks reflect light, which is why egg yolks look opaque. Once you pop the yolk, the disks spring back into spheres. Each tiny ball has the structure of a Tootsie Pop: a core of fat surrounded by a crust of protein, cholesterol, and phospholipids. Fat can add richness to a finished dish, of course, but in the kitchen it's the phospholipids that shine. Molecularly speaking, a phospholipid is a peacemaker. It's got one end that loves water and another that loves fat and functions like the mutual friend of two enemies. Oil and water don't mix, but they both shake hands with different ends of the phospholipid molecule. This process, known as emulsification, means egg yolks can fuse watery and oily ingredients into a smooth whole that can turn, say, butter and lemon juice into a hollandaise.[5]

While the yolk gives fatty flavor and signs peace treaties among opposing molecular camps, the white does a different job. A nearly fatless substance, the white is mostly water but has a dozen different proteins floating in it. These proteins have properties that allow them to foam easily, stabilize foam, become solid when heated, and bind to minerals like iron and copper. Zoom in on the egg white, and you'll see a lagoon filled with balls of yarn. If you expend energy unraveling those balls—think heat or whisking—the unraveled yarn will start to snarl into a solid tangle. But a protein tangle, like gender, is a spectrum. Egg white proteins can go anywhere from completely untangled (liquid raw egg white) to tightly tangled (say, a crispy-fried egg white) and stop at any texture in between. Tangles can also trap other ingredients like air (meringue), fat (custard), and water (fluffy, American-style scrambled eggs). Milk, cream, sugar, and fat from the yolk or elsewhere work on egg white like conditioner on hair; they get in the way of a tight tangle, which translates to delicate textures. According to McGee, salt and acids "get the proteins together sooner, but don't let them get as *close* together." Salted eggs, for example, start to coagulate at a lower heat but don't yield as firm a texture.[6]

Egg cookery presents a paradox. It is the easiest thing in the world—crack an egg into a hot pan, cook, eat—but it is difficult to manipulate with exactness. A true master controls the egg through precise heat, arm power, and additives to create everything from thin crepes to tall souffles. An egg guru is a jazz musician who has mastered the scales in every key and can use them to freestyle a satisfying dish, precisely calibrated to the diner's taste.

Jacques Pépin is my guru of all things cooking but particularly of eggs. For much of my life, I have enjoyed a parasocial relationship with him. He had a cooking show with his daughter, Claudine, that my father and I used to watch together. We devoured everything he produced, from the best cooking show ever made, *Julia and Jacques: Cooking at Home*, which we watched first on PBS and then bought on DVD, to his book *Complete Techniques*, which contains photographed instructions on how to do everything from julienne a carrot to debone a rabbit. I even have copies of some of his old, out-of-print books, including his first one, with Helen McCully, titled *The Other Half of the Egg . . . or 180 Ways to Use Up Extra Yolks or Whites*. Sunday afternoons with my father often consisted of watching Jacques cook a dish, visiting the grocery store, rewatching parts of the show obsessively, then heading to the kitchen to try to replicate the results using our own spin. While our family shorthand refers to many chefs by last name—Wolfert, Mrs. Chang, Bittman—we always called Jacques by his first name in conversation, a rare distinction, shared only by his counterpart Julia (Child), and of course, Kenji (López-Alt). For me, Jacques—or at least his public persona—is a virtual member of the family. He feels like a second father.

I emailed Pépin at the suggestion of a culinary historian I'd been grilling about "fancy" eggs. Surely Jacques would know some elaborate recipes, he'd told me. Jacques is, after all, widely known for his love of eggs. Did I know Jacques? "Yes, he's like a second father," I said. "Oh, so you know him then?" the historian asked. I had to explain that Jacques is my second father in the same way that Lizzo is my inspiration. And that is how I ended up with Jacques's email address.[7]

Imagine my delight when I received a prompt reply from Jacques's assistant to schedule an interview. We put a date on the calendar. I may have spent the next hour yelling to my husband about it. While I was yelling, his assistant emailed me to ask if Jacques and I could chat sooner. "He's been looking at egg recipes, and I think he's looking forward to it," she wrote. I had to lie down on the carpet for a minute, and that was when my phone began to ring. On the other end, a familiar French-accented voice said hello and asked if I could talk right now. Jacques had called me. To talk about eggs. I didn't have my recording gear set up—I didn't even have a question list—but I'd been training for this phone call practically from birth. I pinned the phone to my ear (I will never wash that ear again), opened a Word file on my computer, and began typing.

Jacques has forgotten more about eggs than most of us will ever know. This is the man who taught America how to make French omelets and eggs in cocotte (simmered in little dishes). In the United States, we tend to conflate French food with fancy food. One of the things my dad has always loved about Jacques, though, is that he approaches cuisine more like a craftsman than an artist, in other words, without pretension. Like the egg, Jacques is a culinary shapeshifter; he respects the cuisine of privation—think offal—but he can also whip up a mousseline and coat a trout with aspic. To food lovers like me, he is a standard-bearer for French cuisine, with its fancy connotations, but he is also the guy hired to cook industrial quantities of clam chowder for Howard Johnson.[8]

On the phone he seemed even more energetic than he did on screen, and he was eager to tell me that there are probably three hundred French recipes for eggs, with abundant garnishes, like eggs softly scrambled or cooked in little dishes and garnished with crayfish or sweetbreads. While he was the chef at France's top political household, he sometimes made deep-fried eggs as a first course for dinner parties. He dropped the eggs in hot oil and massaged them with wooden spoons to maintain their shape, then placed them onto toast rounds and adorned them with bacon. Eggs carry the flavor of truffles so well, he said, and you can garnish them with caviar. At the same time, he added, "There is something very democratic about eggs in some way and very unpretentious." During this phone call, and my next one with him, it became clear that eggs are so intertwined with his life that he could trace his own biography in them. So I asked him for a self-portrait drawn through three or four egg dishes, dishes that represent who he is and has been as a person and chef over his lifetime.

While *my* egg biography begins with soft-cooked eggs made

with my dad, for Jacques, it is eggs gratin, made by his mother. Born to a restaurateur and a cabinetmaker in Bourg-en-Bresse, near Lyon, in 1935, Jacques grew up amid bomb blasts targeting the strategic train depot in his town. His father decamped to the mountains to be part of the resistance as the Nazis advanced on the city. His mother stayed in town and fed her three young sons on whatever she could find. She kept chickens, so most of their protein came from eggs, which she hard boiled, sliced, set atop some vegetable—usually spinach or chard—and covered with a béchamel sauce, a French white sauce of butter, flour, milk, and a little cheese to make eggs gratin.

So my dad and I tried a version of the gratin from Jacques's childhood. One of my dad's love languages is sending me the short instructional videos Jacques posts on Facebook. In one of them, jazz chef that he is, Jacques riffs on this childhood specialty, making it simpler and easier for those of us in a rush. In due course, my dad and I piled some baby spinach into the special egg shirring dishes he gave me for Christmas a few years ago and popped them in the oven. Shirring an egg simply means baking it, and the shirring dishes, while nice, are not necessary; a little baking dish or ramekin would do as well. We cracked eggs atop the wilted spinach and returned them to the oven to film over, then drizzled a tablespoon of cream on top and sprinkled them with parmesan—a lazy béchamel. The spinach, cheese, cream, and oozy yolk of the egg blended in our mouths, unctuous and decadent, but less caloric than it sounds and quite satisfying paired with a slice of toast for lunch.

Later, we watched Jacques make a completely different variation on the theme, an unusual staple of his mother's that he has named in her honor: eggs Jeanette. It begins with simple hard-cooked eggs, the yolks scooped out and flavored with garlic, parsley, and milk. Next comes the surprising part. Jacques sautés the filled egg halves in oil and sauces them with the reserved yolk mixture, thinned out with a little water, a dash of vinegar, olive oil, and mustard. My dad and I naturally had to try it—for who has fried a deviled egg? This experiment, with its southern France flavors of parsley and garlic, proved delicious. The extra steps, frying and saucing the eggs, made them feel elegant but didn't add much labor. The egg halves had a slightly crisp, browned surface; the yolks maintained a pillowy texture; and the creamy dressing lightly coated all of it. With a slice of bread and some salad? Heaven. Ever the Renaissance woman, my mother made sure to point out that she had paired our eggs Jeanette with the Christmas carol "Bring a Torch, Jeanette, Isabelle."

Eggs Jeanette might have an ancient antecedent. During a deep dive into culinary history, I came across Martino da Como, the Jacques Pépin of fifteenth-century Rome and quite possibly the world's first celebrity chef. His nickname was "the prince of cooks," and he worked for the pope's chamberlain. He left behind one of the era's few cookbooks, *Libro de Arte Coquinaria*. It's got a chapter on eggs, of course, including some entries I found unusual—eggs poached in sweet wine or milk and topped with cheese; eggs threaded onto a spit; and eggs broken directly onto hot coals. His recipe for "stuffed eggs," though, struck my dad and I as peculiar, so naturally we had to try it. First, we made nonstandard deviled eggs. This filling incorporated raisins pounded to a paste and verjuice, a medieval ingredient that is the sour juice of unripe wine grapes. The recipe also called for aged and fresh cheese (we substituted parmesan and Greek yogurt) as well as "sweet spices," which after a little internet reconnaissance we translated as cinnamon, cloves, and ground ginger. And of course, a meal fit for a papal chamberlain wouldn't be complete without some expensive saffron bloomed in milk. As with eggs Jeanette, we stuffed the eggs, fried them, and sauced them, this time with reserved yolk mixture thinned out with verjuice and heated until thick.

The sweet flavor of the raisins went surprisingly well with the salty umami of the parmesan. But the whole thing tasted like mulled cider—pleasant, although it felt emotionally wrong to eat dessert spices in a deviled egg. Wintry mouth party notwithstanding, I'd rather have eggs Jeanette.

After World War II ended, and at the tender age of thirteen, Jacques left formal schooling and his mother's restaurant to become a chef in the old way, through years of hands-on apprenticeship. He began at Le Grand Hôtel de l'Europe as the kid who stoked the wood-burning fires under the stove, working his way up to different prep stations and through the countryside's fine restaurants, all the way to Le Meurice Hôtel in Paris.

During his time in Paris his draft number came up, and he joined the navy in Algeria. Eventually, someone figured out he could cook and sent him to the Office of the Treasury where he turned out feasts for the top brass. From there, he became personal chef to three heads of state, including Charles de Gaulle. The egg dishes emblematic of this phase in his life, Jacques said, would be poached eggs, or perhaps eggs in aspic, which eventually gave way in Paris to the classic French omelet. I will pen paeans to the French-style omelet in the next chapter, so for now let's focus on poached eggs, both in and out of aspic.

If you search the web for poached eggs, a jillion methods will reveal themselves. I prefer this one: get your hot water shivering, add some vinegar, and crack in an egg. If you want to be fancy and end up with a nicer shape, use Kenji's tip: crack the egg or eggs into a sieve first and let the watery part of the white run off before slipping the yolk and tight white into the water. Gently agitate the top of the water a little with a wooden spoon so the egg doesn't stick to the bottom of the pan. If you are feeling really fancy, per Jacques, you can also use your spoon to encourage the raw white to move closer to the yolk for a more compact shape. Turn off the heat and wait three-ish minutes. Lift the egg with a slotted spoon, and test that the yolk is gushy but the white is set. Since eggs cook fast, carryover cooking—the tendency of hot food to stay hot and to continue cooking when it's off the stove—matters. If not spooning directly onto toast for immediate consumption, stop the cooking by shocking the egg in cold water. A fresh egg always poaches best, and you can estimate how fresh your eggs are by finding the three-digit number stamped on your carton, which corresponds to which of the 365 days of the year your egg was packed. Fresh eggs have tight whites that stand taller and fewer of the loose whites that form spindly wisps in the water, but almost any egg is poachable, if you're not too precious about these things. When the eggs cool, you can clean them up a little with a knife, trimming off those delicate sheets of thin white that spin off them and taste a bit vinegary. I find them delightful, eaten subversively over the bowl of cold water while no one is watching.

What you do with the egg after poaching is up to you. During his apprenticeship, Jacques learned traditional sauces and garnishes for them. For example, you can make eggs Benedict, a dish that dates to the latter half of the nineteenth century (although its exact provenance—New York's Delmonico's, the Waldorf, or a financier's mother—is shrouded in mystery). This classic brunch staple consists of a poached egg placed on a slice of toast topped with ham and hollandaise, a warm sauce made by beating egg yolks with lemon juice and melted butter, and garnished with a slice of truffle. Jacques also learned the old culinary art of aspic. Put the poached egg together with the aspic, and you have *ouefs en gelée*, an elegant and traditional first course. This dish is a runny poached egg, or mollet egg (between soft and hard boiled, with set white but liquid yolk), suspended in a sparklingly clear aspic and decorated with vegetable trimmings. Jacques may have made it for Charles de Gaulle for the Sunday family meal, though the de Gaulle family was into "relatively simple stuff," he said. "I know that I did it certainly for the person before de Gaulle, because the person who was there before was crazy about eggs and truffle." And so I began thinking of this dish as "French President Egg."[9]

The beauty of a French President Egg is that it's fancy and a little fiddly but not particularly difficult. French President Egg uses eggs twice: once in the obvious poached-egg sense and once through the miraculous alchemy that is consommé. My father and I wanted to make the dish in the proper way; we too wanted to eat like we had Jacques cooking for us in Paris. We did the simple part first—we poached some eggs, trimmed them, and set them in the fridge in cold water. The fussier part involved making the aspic, which began with consommé, stock clarified with an egg raft. Our first attempt at consommé failed. I'd tried to cut corners with a pressure cooker stock, and it became cloudy, with fat emulsified into the broth. We ended up with an eggy mess. So we began again. As always, we started with research. Many consommé videos and open cookbooks later, we felt we knew how to do it the very slow and traditional way.

A brown beef stock begins with deeply roasted bones, meat, and veggies, covered with cold water and simmered for half a day. We tended our stock like a newborn baby, stroking it at regular intervals, skimming off the inevitable bone scum and quite a bit of fat. We strained out the solids and began the laborious process of clarification. We took turns pouring it from one pot into another through a cheesecloth-lined sieve to get out as much flotsam as possible. Then we chilled the pan down in a sink of cold water—fast chilling inhibits bacterial growth and eliminates off-flavors—then chucked it on the rear "sun porch," an icy second freezer in Boston's frigid December. With ice-cold broth, it is easy to spoon off the solid fat. At the end of this involved process, we had a nearly fatless, mostly clear stock.

If you have never experienced the divine alchemy of clarifying a stock, I recommend trying it at least once. Time collapses in on itself, moves slowly backward and then forward at light speed. This must be how Jesus felt when he turned water into wine; it's miraculous to watch a dense, disgusting mud yield the purest essence of soup you have ever tasted. Although we'd researched extensively, we decided to use Jacques's instructions for obvious reasons but also because, as my father put it, "he doesn't piss around with heat that is too low." Jacques's method involves using your hands to mix egg whites, chopped vegetables, and sometimes ground meat in a pot, then adding lukewarm stock. As the mixture is heated, the egg white traps particles and carries them to the top where it forms a solid disk known as "the raft." As Jacques explained, the egg white removes some flavor along with the flotsam, so chefs include veggies and meat in the process to return some of it. We poured our lukewarm stock into the icky, eggy slurry and popped the whole thing over brisk heat until it started to bubble. The raft began to form, exactly as Jacques said it would. At first, our formerly clear stock resembled primordial ooze—unappetizing muddy water with chunks of egg-caked veggies and gray wisps of flotsam. Eventually, a disgusting grey-brown scum began to rise and solidify. We stopped stirring, turned the heat to low, and prayed for a miracle. Low heat prevents a strong boil from breaking the raft apart and reintegrating into the stock. Soon enough, the whole thing coagulated into a thick pancake of despair. We let it simmer for the prescribed time.

And then, pure alchemy. We broke the disk with a ladle and

spooned the stock into a cheesecloth-lined strainer. The result stunned us, a stock, now a consommé so clear, my mother said, "you could read a newspaper through it." The recipes all said it would sparkle, and I figured this for poetic exaggeration. But it did sparkle—sunlight reflected off the bottoms of our spoons. We made fairy soup, the essence of soup, what Jacques calls the "ultimate" in his video. It tasted of the raft's bright vegetables against a meaty backdrop.

With the hard work done, we moved to assembly. We added unflavored gelatin to the consommé to help set it up, then poured thin sheets into the bottom of my wedding teacups, which have a nice shape, and chilled them until set. We blanched veggie trimmings to help them keep their color—leek shreds, minuscule carrot wedges, and parsley leaves for decoration. Night had fallen, and my mother and husband joined us in the kitchen. We took the trimmings, lamenting our lack of culinary tweezers, and formed tiny flowerscapes in the teacups. My dad and husband each did one or two, but my mom really got into it, taking her time, rendering several beautiful botanic landscapes that combined both her love of green things—as she once told me, while buying a flat of plants, "To love plants so much is a terrible sickness"—as well as her fascination with cute decorative stuff, which I share. We placed our cold eggs on the designs, topped up the teacups with the gelatinized consommé, and slid them into the fridge to firm up for New Year's Eve dinner, an extravaganza that would feature a complex omelet (an omelet Arnold Bennett of smoked haddock, hollandaise, and cheese) in addition to this eggy first course.

A few hours later, after we'd fed my son—he lapped up a bowl of that gorgeous consommé—and put him to bed, we pulled the teacups out of the fridge and, after a quick dip in hot water, inverted the eggs onto toasts. They were the most beautiful thing I have ever eaten—they looked like antique paperweights of brown-tinted glass, the egg a pale background for the flowers frozen inside. They also tasted like nothing I have ever eaten. The liquid yolks encased the slippery meat jelly, the two delicate textures and flavors coming together in mouth-coating umami. After a few bites, though, our plates looked a mess; the paper-weights rapidly disintegrated. Perhaps we should have used more gelatin, my father said, they might have held together better. The only other way to improve them, he added, was a set of dedi-cated aspic molds. Did I need a set of aspic molds? I could see my husband begin to stiffen up. I'll get you a set of aspic molds, my dad said, if it doesn't ruin your marriage.

Two weeks later a set of four arrived in the mail.

I've made French President Eggs a few times since with the consommé left in my freezer, for friends I knew would be game and appreciative. I made them for myself too, a light little lunch on toast with a salad. But the labor of consommé deterred me. Jacques told me that his wife adored this dish and that they'd eat it five or ten times every summer. He also confessed, "In the summer, I don't even clarify the stock if I have a good chicken stock." (After I told my dad what Jacques said, he bought his own set of aspic molds and started cheating with store-bought consommé from a can, which he claims is almost as delicious.)

Part of the reason I love French food in general and Jacques's cookery in particular has to do with my mother's tastes. After radiation damaged her salivary glands, she's been unable to eat anything too salty, spicy, or acidic—she used to find orange juice too much for her palate—but many French flavors are safe for her. My devotion also has to do with broader cultural trends. Today's tastes, Kenji told me, have "a lot to do with our asso-ciation with French haute cuisine—our association with fancy food being French or European, so I don't know, it might change with the times but it's hard to dissociate from the time." He also pointed out that in fine dining "we tend to associate delicate flavors with fancier better food—it's a cultural imperialism thing." French sociologist Pierre Bourdieu might agree. In his influential book *Distinction: A Social Critique of the Judgement of Taste*, he makes the case that our aesthetic choices—everything from what we eat to what we read and wear—says something about our social class. For example, fancy East Coast elites eat arugula while "real Americans" have juicy burgers and domestic beer.[10]

But in themselves, eggs don't carry class associations. As Jacques told me, they are "probably one of the most democratic foods because even in a small country where people have no money there tend to be a couple of chickens running around." The food historians I interviewed agreed. Yale professor Paul Freedman, author of *American Cuisine and How It Got This Way*, told me they are "one of the few foods that everybody eats," while Ken Albala, author of *Three World Cuisines: Italian, Mexican, Chinese* and professor at the University of the Pacific, explained that eggs are "among very few foods that don't have any class associations at all." A full English or Scottish breakfast—a huge spread that includes fried eggs—has working-class connotations, he said. Meanwhile, King Louis XIV of France had a crier proclaim, "The King is about to eat his egg," as a crowd gathered to watch him lop the top of his soft-boiled specimen. People of all classes eat eggs, and whether they are plain or fancy lies in the treatment and cultural context. Jacques's mother hard-boiled them in war time and covered them in a simple cheese sauce—rustic for him, but the use of the word *béchamel* in the United States adds a touch of foreign elegance.[11]

It all makes me wonder: do my father and I love French egg recipes for themselves? And to what extent is our taste the product of my upwardly mobile dad's childhood during the height of

French-food mania in the United States? Does any of this matter if we enjoy the process and the meal? The only thing I know for sure is that making and eating French President Eggs is a unique experience to share with friends and loved ones, a little novelty, an aesthetic jolt to the eyes and tongue. And for me, that's enough.

VELVET EGGS

I enjoy many egg dishes, but, as I noted ear-lier, I prefer a single egg texture—custardy. Soft, a little goo-ey, bordering on runny, velvet on the tongue. For me, a classic French omelet—a thin skin of cooked egg wrapped around a custardy scramble, egg containing egg neatly in its own self-en-velope—is the ultimate egg dish. Though the egg man himself, Jacques Pépin, once told me he was in the mood for a "large curd country omelet for dinner" that night, the classic French omelet is still his favorite preparation.

The question of how to make an ideal omelet is one for the ages. I can find little to improve on in the general guidelines set forth in *Le Ménagier de Paris* (ca. 1393), perhaps the first French omelet recipe. Written in the voice of an older gentleman educat-ing his teen bride, the book is proof that mansplaining knows no historical boundaries. This recipe for a pair of omelets uses an astonishing quantity of herbs—thirteen different ones mixed to produce two handfuls. The cook is to grind them with ginger, then beat that mixture into sixteen—sixteen!—eggs, formed into two omelets:

First, heat the pan thoroughly with oil, butter, or another grease as you wish, and when it is good and hot all over,

especially toward the handle, mix and pour your eggs into
the pan and turn them often with a spatula, then sprinkle
on some good grated cheese . . . then cover with the edge
of the eggs.[1]

In other words, get that oiled pan hot, pour the beaten egg
and herb mixture into the pan, and move it around. Sprinkle
on cheese (if using) and fold. That's it. And if we leave aside
questions of terminology, like whether the solid egg-and-garnish
disks of kuku, frittata, or Spanish tortilla are omelets—this is
still the basic recipe for pretty much all omelets. But what oceans
of subtlety lie wrapped up in those simple directions!

Get the pan hot—but how hot exactly, and what kind of pan?
Add the butter or oil—but again, how much? What does "turn
often" mean? Is this a recipe for the classic French envelope of
scrambled eggs, a large-curd country omelet, or a simply folded
egg pancake?

Believe me when I say that over the last quarter century my
dad and I have made omelets with the best of them. We have
tried one-, two-, three-, even eight-egg omelets. We have tried
twelve different kinds of pans. We have tried many different
garnishes, from cheese and herbs to smoked haddock and hollandaise (omelet Arnold Bennett). We have tried gas flame and
electric heat, blazing pans yanked on and off the burner, as well
as cooler ones resting over a low flame, eggs beaten with fork,
chopsticks, whisk, and spatula. We have tried plain eggs, eggs
with salt, eggs with salt and tabasco, eggs diluted with water
and milk, even herbs and fried bread crumbs beaten right into
the mixture.

We still cannot make a perfect classic French omelet every
time, and frankly that is part of the omelet's appeal. My dad
explains why he keeps on trying: "My interest in omelets grew

because they are hard to make correctly. If I could make one right every time, I'd probably be less interested in them. But I have tunnel vision and a little bit of a perfection neurosis." Only a master can make one perfectly every time, and Jacques is a master, undoubtedly. And though we have watched his nimble hands turn out omelet after omelet, picked up on this or that aspect of his technique, it is still no substitute for practice. I can only conclude that Jacques has brains in his hands, a chef's finely honed sense of exactly how eggs behave.

And yet the precariousness of the omelet, the fact that we get it perfect only one out of every five times, adds to the charm. As my dad says, "What you are doing is ephemeral—you make an omelet, and you eat it and then what you have is the memory. You can't put it under glass and say I did that; you have to say I ate that." Omelets are evanescent; to make a fine one is to master the variables of heat, arm motions, and additives, and play them against one another in exactly the right manner. Each omelet is its own little art happening—here today and gone in a gulp.

While I have many memories of making omelets, I have few of eating them. Everything runs together in one bite of buttery, eggy goodness. After I moved out of the house, though, I used my omelet pan—the trusty carbon steel one my dad gifted me before college—to run workshops for my friends, and these, plus the lessons from my dad, are what I remember. Psychologist Mihaly Csikszentmihalyi first described the concept of flow in the book of the same name, and for me, an omelet workshop produces many of its hallmarks. A person achieves flow when immersed in a task that is challenging but not beyond their skill, intrinsically reward-ing, and structured to make progress visible. All of these describe omelet making. Achieving perfection is tricky, practice inevitably nets improvement, and almost all results taste delicious. Improve-ment comes from keen attention to all the senses: touch—the fork that jumbles up the egg as soon as it hits the heat, the weight of the pan you jerk back and forth until the crucial decision to let it set; sound—the egg's sizzle in hot but not too hot butter, your friends yelling advice at you; sight—the pan you jerk back and forth until the crucial decision to let it set when you can see the curds are just so; smell—buttery, not browned; and taste—rich and custardy.[2]

During one workshop in a vacation rental, perhaps five of us crowded around two pans on the stove, feeding our successes and failures to the crowd on the porch. Two hours and four dozen eggs later, we stopped because my friend, culinary experience designer Avital Ungar, made the perfect French omelet. We stared at it with pleasure, congratulating her on the pale, tender, buttery goodness. The chefs saved that one for themselves.

My younger cousin Kitt, who loved to cook, lived with my parents for his last year of high school. He also received omelet lessons from my dad—a sort of brotherhood of the pan. My dad, of course, gave him a carbon steel skillet as a parting gift for college. When I was in my early thirties, Kitt, then nineteen, skied into a tree and died. Our family was beyond devastated. A few afternoons after the tragedy, some friends came over to keep me company. It felt so natural to queue up a series of instructional videos, retreat to the kitchen, and work through a stack of eggs. I remember laughing together, feeling Kitt's presence among us, as my husband came home with supplementary supplies. One of us held a golden mandorla on a plate, and all four of us reached forks forward simultaneously.

My practice, but not perfection, of the classic French omelet has convinced me that those who master the egg are magical. It is easy to mistake equipment for skill, my dad explains, and that is why he has a veritable museum of omelet pans, more than a dozen. On a video call, he spreads them all out for me: three reproductions of Julia Child's favored pan, made of old torpedo shell ends; the aluminum Mr. Omelet pan he and his sisters gave their father in the 1960s; a bookend of cast-iron pans, the first purchased for himself as a young college student, the second a special antique pan with a smooth milled bottom, purchased on retirement from tax lawyering; and what, by his accounting, was the holy grail of pans—blue French steel, which is really a romantic way of saying carbon steel that stains. Cheap and indestructible, they are the traditional choice for making a classic French omelet. Like cast-iron, blue steel pans require seasoning—the gradual buildup of a patina of oil, polymerized by heat. I have the twin to his eight-inch carbon steel pan in my cupboard, with my own slowly growing museum of omelet gear.

My mother wandered into the frame, and before she could see the kitchen, my father placated her with "I'll put all those pans back when I finish talking to Lizzie." She muttered something about needing a blindfold instead of a face mask, but after forty years of marriage, she's resigned to the culinary overkill. Some men retire and start gambling or buying sportscars, she once told me, but all my dad wants is omelet pans. And trumpets, but that's for another book.

Why so many omelet pans? It has to do with the proximity to greatness. Take Jacques, my dad says. "You look at the equipment that he is using and say, 'I wanna be like him and all I need is that pan,' when in fact he could probably do the same thing with a stick over the fire. So you can be kind of an omelet pan-o-holic because it's really hard to get an omelet just right. You put it on, and unless you've done it so much that you speak the language of eggs and can hear what they are saying, you're fucked. So if you can't speak omelet, can't have a dialogue with the egg, the next best thing is to go and get the next pan." A poor carpenter blames his tools and all that, but sick gear is its own inducement.

When I talked to Jacques about omelet pans, he told me that in Paris he worked the breakfast shift for a while, and each night he hid his omelet pan. "I took it when I came in in the morning, and just cleaned it up by wiping and put it in my closet," he said, the closet being his dressing closet at the restaurant. I remember my own feelings about that omelet pan in my college days. I contributed a recipe to the family cookbook that included the phrase "your special omelet pan, which your roommates must never touch." Jacques, of course, was safeguarding his work. If you leave a pan like that in the kitchen overnight, someone might boil water in it and ruin your next breakfast shift. "I have one of those old pans still on my wall here, and it's [got] a beautiful sway, it's not a deep corner—it's a really beautiful shape on top. It's on the wall here. I don't use it because it sticks. Why does it stick? Because I don't use it." Jacques has hung up his carbon steel pan. We are not cooking like Jacques anymore.[3]

That's when Jacques also discussed the large curd country omelet. He told me, "Basically that's how my mother and my aunt would make an omelet," later adding that "most people in France are not going to do a classic French omelet." I had thought I was emulating French home cooks, but the omelets that my dad and I have been attempting to perfect all these years are not the ones that French home cooks make. The classic, Jacques said, is more for fine dining, which is why, for him, it is the egg that represents his period in fine restaurants in Paris.

I do more homework on the classic French omelet before I talk to Kenji. I am insistent, zealous even, about cooking an omelet in blue steel. Kenji prefers a nonstick pan and tells me, "If you're using carbon steel, you really have to get the temperature exactly right." That I knew: the difference between a seasoned pan that freely releases tender egg and one that does not or, worse, browns the omelet, may be a millimeter on the stove dial, two minutes of preheating time, the temperature of one's eggs, or all three. An omelet is finicky. From all this I deduce two things: cooking the classic French omelet is an expert-level skill, and you better be a master if you are cooking one on a seasoned steel skillet, because there is zero room for error.

All these years we've been trying to cook an expert dish on hard mode.

I have a vendetta against nonstick. When I was sixteen, a scratched nonstick pan imparted the flavor of old salmon oil into an otherwise perfect omelet. For years, I have been saying that they are bullshit, part of the American movement of planned obsolescence. Nonstick pans scratch, but carbon steel is forever. And yet. Nonstick technology has apparently advanced over the past few decades, and it was the choice of pan for the two chefs whose advice I venerate. Kenji makes omelets on nonstick, and so does Jacques. Who are we to argue?

So my dad and I added to our omelet pan collections yet again. And I confess I have done a one-eighty. I was attached perhaps to the romance and sheer difficulty of a steel pan, but the nonstick is foolproof. I simply changed out my omelet-making fork for some wooden chopsticks, and whether the heat is super low, low, or medium low, voilà! I turn out perfectly creamy and pale specimens every time.

I cannot get enough of the classic French omelet. But I don't prefer it in a classic way, at least not a French classic way, with cheese or herbs. I prefer it as made by the gang of vagabond chefs in the classic 1985 food film *Tampopo*, which centers on the relationship between people and food in modern Japan. This delightful film takes its inspiration from Westerns and kung fu movies, so naturally it's about a cowboy truck driver who decides to train a single mom to become the best ramen chef in the land. Cue the training montage, which involves lifting a lot of stock pots. Its B story involves a gangster and his girlfriend's erotic explorations with food: a memorable scene includes them getting intimate while passing a raw egg yolk between their lips. It also has a few short segments loosely based around food themes. In one of them, the cowboy takes our heroine to meet the gang of vagabond chefs who hang out behind the city's best restaurants and finish all the fancy, half-drunk wine bottles. While she's there learning about wine, one of the vagabonds absconds with her kid to make him a snack. Silently, to Charlie Chaplin music, they sneak into a commercial kitchen while avoiding its security guards.

The boy's mouthwatering prize is a perfect French omelet perched atop a mound of rice that has been stir-fried with ketchup. The vagabond slices down the middle of the omelet to reveal its gooey interior, then tops it with more ketchup. It glistens with promise. And since Japan is very far away, and more elaborate versions of this dish (called omurice) on YouTube were years off, my father and I rewatched the scene nine or ten times and headed to the stove. This is still my very favorite omelet. The jagged texture and tang of the ketchup rice play against the eggs' creaminess. The added ketchup on top has its own different pudding texture, lighter and thinner than the unctuous egg. Tomato and eggs are, after all, a classic combo, both in French cuisine and in Chinese cuisine, where stir-fried tomatoes and eggs are among the first dishes kids learn to cook. I am fond of the Chinese fry-up, but nothing can replace omurice in my heart. Part of the romance lies in the flavors and textures, but it also lies in my memories of watching that niche film with my dad, then using our wits to concoct our own version.

My dad's obsession with the perfect omelet comes from his mother, the farm girl I mentioned who raised chickens. She was meticulous and insisted that everything get done the right way, aka her way, no matter how long it took. It also comes from my dad's father, who grew up the eldest of seven children in

an impoverished upstate New York home. He went to technical college on the GI Bill—the first in his family to finish this
level of schooling—and became an engineer who designed drone
planes to fly around H-bomb tests taking measurements. For
my grandpa, who grew up in poverty, food, particularly fancy
food, could be a supreme adventure. One year for Christmas he
wanted to try roast suckling pig but famously brought home an
enormous yearling, frozen, with legs extending fore and aft. It
required a hacksaw to get it into the oven. He also made real
Danish pastry with my dad after reading about it in a glossy
magazine and once organized a shipment of live lobsters from
Maine with some of his colleagues at the air base. My dad brings
his father's sense of adventure to our investigations, as well as
his mother's sense of precision. And he has imbued that sense of
adventure in me, which has brought us our eggy obsession.

Before we leave the culinary uses of eggs entirely, I want to discuss the last of Jacques Pépin's biographical eggs. In his eighties,
Jacques still loves eggs and delights in new ways to cook them.
As he told me, "The most recent way I'm doing eggs would be
the eggs cooked *sous vide*—so cook it at low temperature, well,
148, 150 [Fahrenheit] for an hour, and it does a beautiful texture—the egg yolk and egg white is a beautiful texture." French
for "under vacuum," *sous vide* involves cooking a food for a long
time in a water bath set at a precise temperature. Although it's
often billed as a new form of cookery, Jacques points out that he
used the technique of "cryovacs" as a chef at Howard Johnson's
in the early 1960s—"basically the same thing," he said. Typically, one uses a *sous vide* bag—a plastic bag with the air sucked
out—but eggs, of course, come in their own air-tight packaging,

so it is possible to drop one into a water bath and wait an eternity in egg time (forty-five minutes) to uncover new textures. One can pasteurize eggs while leaving them basically raw, create unbelievably creamy soft-boiled textures, or if one heats them to Jacques's preferred zones, create an egg that ranges from perfectly poached in the shell to one with a fudgy, malleable yolk, a texture virtually impossible through other techniques. An added advantage is that the method requires little attention and is foolproof. The results, straightforward but luscious, go with the aesthetic of a chef whose recent books include titles like *Quick and Simple.*[4]

When my spouse and I first brought home a *sous vide* circulator, we immediately began testing eggs in it, per the famous egg chart of Dave Arnold, director of culinary technology at the International Culinary Center in New York. That chart, reproduced broadly around the web, moves from soft-cooked, through the strange *sous vide* midground where yolks remain malleable and can be rolled into sheets, all the way up to hard boiled. (The consensus, though, is that if you're hard-boiling an egg, it's best to do it in a pan because it will peel better.) I love a *sous vide* egg, and if it weren't for my friend Jeeyon Shim, a game designer, that might be the last egg in my own biography, behind soft-boiled and omurice.[5]

However, Jeeyon sent me a recipe for a favorite comfort food of hers: ttukbaegi gyeranjjim, a Korean dish of eggs steamed in an earthenware bowl. The resulting soft eggs, eaten with a spoon, reminded her of her mom. Steamed eggs out of the shell? I'd never heard of it, but I was intrigued.

Steamed eggs, apparently, are quite common in East Asian cuisine; different countries have their own versions. One mixes eggs with liquid and then sets them up by steaming them in a ramekin in or over simmering water or by gentle cooking in a lidded container. The ratios of eggs to liquid vary widely, from two parts egg to one part water all the way up to one part egg to three parts water. As my father's daughter, I started with the simplest variation I could find, which is the basic recipe for Chinese steamed eggs, or "water eggs." Although fancier versions can include ground pork or other fillings, the basic version contains only eggs, water, and a pinch of salt. Following Wei Guo's recipe "Chinese Steamed Eggs, A Perfectionist's Guide" on the blog Red House Spice, I mixed one part egg with two parts water and a pinch of salt, then poured them through a sieve to remove any lumps. Different chefs have different techniques for keeping the surface pristine, as steam that condenses on the pot lid can drop down and mar the eggs' surface. I followed Wei's directions, placing cling film over my filled ramekins, which I then punched holes through so the steam could escape. I set the ramekins in my vegetable steamer and heated them until set: since I'd mixed them with cold water instead of Wei's suggested warm water, it took much longer than the recipe's ten to twelve minutes. I removed the little custards, scored their tops in a decorative manner, and drizzled them with light soy sauce and sesame oil, then scattered a few chopped scallions on top. The dish blew my mind. It had a texture somewhere between flan and crème brûlée—silk on the tongue but with a pleasant needle of salt from the soy. It was light, silky, smooth, slippery—everything I love in an egg. I became a woman obsessed, an obsession I soon transferred to my father, who put out an immediate order for kitchen gear—lidded teacups and a supersized steamer to hold four at once—after his first water egg hooked him.[6]

Of course, we couldn't stop after trying only one recipe.

In Korea's gyeranjjim, broth, traditionally anchovy broth, replaces the water, and the mixture can also include things like dashi, mirin, and sesame oil. The method of cooking varies. Jeeyon steams hers in a bowl atop water, but there are other methods. One can heat the broth in a stone bowl called a *dolsot*, then swirl in beaten sieved eggs as well as fillings like shredded carrot and scallion. Or the fillings may be added when the eggs are partially cooked but still runny, so anything particulate ends up evenly suspended through the finished dish. The chef covers the *dolsot* and lowers the heat, and the eggs puff up as they cook. Proportions vary but tend to be closer to a 1:1 ratio of egg to liquid. Gyeranjjim is a succulent mixture that is part milk-less custard and part soup. Due to the lower water ratio, the texture of the egg wasn't as delicate and wobbly as my first water eggs. The cooking method in the covered *dolsot* set directly over the heat also created a different texture—the water eggs had the smoothness of soft tofu, while gyeranjjim was pillowy, with visible curds—plus there were a few tablespoons of savory broth in the bottom to slurp up. I found the results homey, delicious, and comforting.

In Japan, chawanmushi uses dashi for its liquid, often seasoned with mirin and light soy sauce. As with all of the steamed eggs, egg-to-liquid ratios vary, but here they range all the way up to one part egg to three parts liquid, an extreme that made our final dish the wobbliest of all. For garnishes, think cute and artsy—a single shrimp, a slice of mushroom, and a carrot cut into an ornamental shape. My father and I may have overdone the garnishes (or so proclaimed my mother-in-law, who lived in Japan for a couple years), but the final product still proved delicious. Chawanmushi presented a gorgeous way to extend the flavor of a few shrimp. And straining the mixture into teacups, as we did with the Chinese eggs, gave the dish a silky smooth texture. The higher liquid-to-egg ratio, though, left us with some mouthwatering broth in the bottom of each cup, more like our try at gyeranjjim.

As I documented my steamed-egg obsession on social media, friends commented that their grandmothers cooked similar dishes for them; steamed eggs meant comfort or represented a favorite breakfast. In a word, I had made my own, perfect Dick Stark experiment. We tried something new to us, which was simple, healthy, and satisfying. It had the benefit of sustaining infinite variations as well as discussions of technique. We spent quality time together in the kitchen, talking about my kid, my writing, and my dad's trumpeting while we worked. And, of course, the dish ticked all the best "mythic" boxes—epic ingredient, new-to-us preparation, opportunity for new gear, and best of all, the sacred image of Grandma making comfort food for the people she loves. In the absence of a Chinese, Korean, or Japanese grandmother, we could cook like one, entertain the flavorful results on our tongues, and wonder about the fabric of life in cultures not our own.

Food goes deep as a vehicle for personal identity. Jacques's biography in eggs is gratin, poached, omelet, *sous vide*. For me it is soft-boiled, omurice, *sous vide*, and now, steamed. My father makes steamed eggs four at a time and pulls the extras out of the fridge to have with toast, doing what people always do with an egg dish—changing the parameters to fit their exact taste. Egg, the culinary shapeshifter, builds bridges across continents and cultures because it is versatile enough to suit many palates. Perhaps we are not unique snowflakes but precisely calibrated egg dishes. I am reminded of the paradox of eggs again: you can put anything in them, and they will taste good. As Jacques sometimes says in his videos, if you mess up, no one will come and arrest you. The stakes are low, but the flavor is high. And like an egg on the hob, getting the best out of a person means treating them with care, treating them just so, maybe even treating them to their favorite egg.

7

PYSANKY

Many world cultures love a decorated egg. In China red eggs symbolize luck and are eaten on birthdays—especially a child's first birthday—and on other special days, such as a child's first day of school. Nowruz, the Persian new year celebration, includes eggs decorated using a variety of techniques. They may be hard boiled in solutions tinted with beets (pink), red cabbage (blue), turmeric (yellow), and other natural dyes. Sometimes people tie herbs or flowers to the eggs, which they wrap in cheesecloth and then boil so that the shells exit the pot with a lovely reverse image on them. People also paint or draw on the eggs with markers. I've seen Nowruz eggs with pictures of fruit, people, goldfish, calligraphy, and abstract designs. The Persian wedding tradition of *sofreh aghd*, a ceremonial spread, also includes decorated eggs that may be encrusted in lavish materials, such as silver beads, rhinestones, or gold leaf, and represent fertility for the couple. In Mexico, *cascarones*—decorated eggs that have been hollowed out and filled with confetti—make joyful explosions as they are broken on children's heads in springtime. Adolescents, too, have been known to flirtatiously break them on one another. In a folk tradition from Hungary, meanwhile, blacksmiths display the delicacy of their skills by decorating blown eggs with miniature horseshoes.[1]

Decorated eggs have a long history as art objects. Consider finds at Spitzkloof A, a prehistoric cave dwelling some sixty thousand years old in the midst of South Africa's Richtersveld desert. There, buried beneath layers of ancient silt, archaeologists found fragments of ostrich eggshells, unsurprising given that ostriches lived all over Stone Age Africa. But these eggshell fragments came in a rainbow of hues—bright teals and reds, soft browns and buff colors, white and black. The shards got dig director Dr. Brian Stewart wondering whether the colors resulted from purposeful human intervention or environmental exposure. The answer wasn't obvious. After all, these eggshells had spent about sixty millennia underground, in a part of the desert where ancient humans built plenty of fires.

Dr. Stewart couldn't wait around for sixty thousand years to find an answer, so he decided to play with fire in an informal experiment with his team. First, they dug a large hole. Next, they buried broken ostrich eggshells at varying depths. Finally, they built and tended a hot wood fire for hours to mimic presumed early human behavior. When the area cooled down, they dug up the shell fragments to see what had happened to them. The pieces on the surface of the dirt, exposed to the most fire, turned white or black. A little bit below the surface, the shells had turned deep brown; then below those, bright yellow; and finally, at the bottom of the stack, a milky off-white. Although informally done, the experiment failed to turn any shells bright teal or fiery red, leaving open the possibility that people had dyed at least some of the eggs.

It made sense to find decorated shells here, given the rest of the archaeological record. About 500 kilometers (310 miles) to the south of Spitzkloof A lies the site of Diepkloof, a rock shelter at least sixty thousand years old. Over the years, archaeologists working there have also found ostrich shells, including fragments that have been carved with patterns that resemble railroad tracks, checkerboards, or parallel lines crossed at oblique angles. As Dr. Stewart wrote, these jagged bits of shell are "among the world's earliest evidence for abstract thought, and thus ways of behaving that resemble our own."[2]

The decorated eggshells had a purpose too. Dr. Stewart theorizes that the designs showed ownership of individual eggs as well as membership in a group, the same way some people put bumper stickers on their cars or decals on their laptops.

Early humans used ostrich eggshells as canteens. The eggshells are durable, and their sturdy construction forms an architecturally sound dome. As Dr. Stewart explains, "They're robust, not too large—(1 litre on average) and breathable, so water stays cool. . . . It is probable that the innovation of using ostrich eggshells as water flasks allowed humans to live in Sub-Saharan Africa's driest areas." Obtaining these eggs was dangerous work, as ostriches can grow up to 9 feet tall, weigh 330 pounds, and grow claws 4 inches long. Capable of a top speed of 45 miles per hour, their powerful legs can deliver kicks strong enough to kill a lion. Fortunately, ostrich flocks use a day-care model for egg supervision, laying all their eggs—as many as sixty—in a communal nest, which is a pit in the sand. The most dominant female lays her eggs in the center, while other females lay theirs at the periphery. The dominant female can tell which eggs are hers and rolls some of the other eggs away, since one bird can only incubate twenty or so at a time. The most dominant male and female take turns incubating and guarding the nest. The pale, gray-brown feathers of a female ostrich allow her to camouflage the eggs during the day; the male's black ones do the same at night. A few times a day, the ostriches stick their heads in the nest to rotate the eggs, which gave rise to the famous head-in-sand myth. Any predator that can defeat the guardians has a cache of eggs for the taking.[3]

Ancient humans, after circumventing the ostriches, stole the eggs and prepared them to hold water. For his how-to guide, Dr. Stewart looks to the San peoples, Indigenous inhabitants of the Kalahari, who still use egg canteens. The San cut a small hole into the top of the egg, then insert a strand of grass or a reed into the shell and whip up yolk and white so they will pour out easily. An ostrich egg is the world's largest cell, equivalent in volume to about two dozen chicken eggs, and the San don't waste its contents. They pour the whipped egg onto a hot surface to make a huge omelet. The empty shell, washed and fitted with a grass or beeswax stopper, then becomes a canteen. Early humans also made use of the broken shells. They ground them into tiny beads and exchanged them across social networks spanning hundreds of miles.

Other cultures used ostrich eggs symbolically. For the ancient Egyptians, ostriches were associated with the Sun since they reputedly flap their wings and run about early in the morning. Maat, the goddess of death, truth, and justice, wore an ostrich egg in her cap, and her symbol was an ostrich feather. In the afterlife, she determined a person's morality by weighing their heart on a scale against that feather, the feather of truth. The fate of souls hung in the balance, as bad deeds made a heart heavier. If the heart weighed more than Maat's feather, the monster Ammit—a crocodile-lion-hippo hybrid—devoured it immediately, but if the heart weighed the same as or less than the feather, the dead person could continue on the path to Osiris's eternal paradise. The Egyptians painted or scratched designs into ostrich shells, buried them in tombs, and offered them to pharaohs as tribute.[4] To the ancient Greeks, ostrich eggs symbolized fertility and prosperity. Archaeologists have found shell fragments in temples dedicated to Apollo as well as to the twins Castor and Pollux, who were born from an egg after Zeus, in swan form, had sex with Leda. Ancient Greeks traded decorated ostrich eggs—carved, painted, or embellished with metal—as luxury goods throughout the ancient Mediterranean. For Coptic Christians, meanwhile, ostrich eggs represent both womb and tomb. Chicks burst out of eggs like Jesus burst out of Mary's virgin womb and, later, his tomb. In addition, ostriches were once believed to hatch eggs by staring at them continuously and thus represent vigilance and focus, qualities that worshippers should direct toward God.[5]

Because of the egg's various meanings—as symbol of purity, rebirth, vigilance, and faith—Italians hung them in churches. Famously, Italian Renaissance master Piero Della Francesca's Brera Altarpiece shows the Madonna and child surrounded by saints and angels, with an egg suspended from the dome in the ceiling. Art critics have likened the egg in the dome to a pearl inside an oyster, the metaphorical grit of the flawed believer transformed through God's love into perfect new life. That painting gave rise to one of art history's more internecine scholarly disputes from the 1950s to the 1970s, a multidecade issuing of papers about whether the egg is an ostrich egg or not. Scholars tried to reverse-engineer the size of the egg based on Francesca's precise use of perspective, argued about whether he would ever have seen an ostrich egg (is this a nod to the ostrich-owning patron of the painting?), and made vigorous symbolic arguments. Perhaps, one scholar opined, it is actually Leda's swan egg. In this case, the Madonna and child in the foreground could represent the new religion blotting out the old. For me, the egg in the painting conjures all of it—the essence of divine birth that symbolically mirrors Mary's birth of Jesus, the rebirth of a Christian in God, and the beginning of a new faith, with all the devout vigilance that implies.[6]

With so much rich symbolism embedded in the egg, it's no surprise that humans have both decorated them lavishly and developed complex customs around them. Nowhere are the affordances of egg decoration on more thorough display than in Venetia Newall's 1971 monograph *An Egg at Easter: A Folklore Study*, which explores egg customs throughout the world and is something of a laundry list of all the weird things humans have believed about or done with eggs over time. Newall links eggs to concepts of fertility (obviously), purity (nice white shells), resurrection (life from a stone), and sacrifice (eggs often replace live animal sacrifice), among others. Decorated eggs have been buried in the foundations of buildings, cracked on the stomachs of women in labor, given or not given to young women out of a belief that they enhance or destroy fertility, eaten as an aphrodisiac, buried with the dead, and used to summon spirits, to name a few. Newall also records an incredible array of springtime folk traditions. In Eastern Europe, decorated eggs were buried in the fields, for example, to secure a good harvest or exchanged among young people as signs of favor. While springtime egg decoration probably pre-dated Christianity by a fair clip, Christianity wrapped the tradition of decorating eggs into its own mythology. An old Polish legend says the Virgin Mary colored eggs to please baby Jesus, while another legend declares that Mary Magdalene took boiled eggs to Jesus's tomb for lunch while she anointed his body and that the eggs turned bright colors—a miracle!—as soon as she arrived.[7]

At any rate, only religious syncretism can explain the many customs related to Easter eggs. For starters, they were a curative that could treat, as Newall puts it, "every imaginable complaint" from lumbago to bad skin. But the wildest custom Newall writes about—by far—is this one:

A man with religious connections, Adam Weikand, personal physician to the Bishop of Fulda, described a secular, peculiarly disagreeable ceremony, which he saw in the late eighteenth century. A live donkey walked in the Palm Sunday procession, and women there placed colored eggs inside its anus, so that they could be blessed.

While I tend to doubt stories that resemble urban legends, I can't help but imagine a group of German peasant women, gathered around the donkey before the church service, saying to one another, "We're about to do Christianity, right?"[8]

The social customs around Easter eggs fascinate me. Many countries have traditions of begging for eggs. In Sweden, for example, witches are supposed to meet up with the Devil for an orgy on Maundy Thursday—gotta love the Nordics—so the children dress up like "Easter crones," or "Easter witches" (*påsk-kärring*), wearing old, oversized clothing, rouging their cheeks, and painting on freckles. Then they trick or treat at neighbors' houses for decorated eggs and candy. In rural Hungary and elsewhere, boys douse girls with water on Easter Monday, also known as "wet Monday." The girls can avoid a second ritual bath by paying a ransom of decorated eggs. Slovenian girls once gifted red eggs to their heartthrobs, hoping to secure love in return. Poland, meanwhile, once had a tradition of auctioning girls off. The girls sat up in a hayloft while their sweethearts bargained for them as for a cow or horse, mentioning all their finer points. The winner received a basket of eggs the young woman had decorated.[9]

But nothing rivals the Eastern European tradition of making and exchanging pysanky, an ancient art of egg decoration. Every place has its own unique designs and techniques. In Poland, for example, *pisanki* include *oklejanki*, eggs decorated with collages of yarn and the spongy interior of reeds, and *pacenka*, eggs dipped in wax and then scratched and dyed. In Ukraine, which is famous for the art, pysanky features delicate and intricate patterns. The designs resemble those of quilts and embroidery, with ladders, grids, rakes, triangles, stars, whorls, zigzags, stalks of wheat, leaves, and other motifs in eye-popping color combinations. Passed down from mother to daughter, the design and color schemes have their own special meanings that vary from village to village and practitioner to practitioner.

Tracing the exact contours of the pysanky tradition is difficult, as the folk art nearly died out during the Soviet rule of Ukraine, when authorities forbade it due to its religious affiliations. In addition, Soviet authorities purposely destroyed historic pysanky collections, while others were lost in the wreckage of World War II. In *The Ukrainian Folk Pysanka,* author Vira Manko explains that "during the era of Soviet occupation in Ukraine, this form of applied art was subject to total destruction." Ukrainian American pysanky artist and researcher Luba Petrusha echoes that thought on her very thorough website about the history of the art from the 1930s to the 1980s: "It was, truly, as though the pysanky had never even existed in Ukraine. It was not studied. It disappeared from museums, and there were no books or exhibits on the subject." But pysanky finds a way. Petrusha continues, "There were individual women who, in their villages, preserved traditional Ukrainian crafts and rituals, writing pysanky during the Easter holidays, keeping the old traditions and designs alive." According to one Ukrainian folk legend, the fate of the world rests on pysanky, as the eggs strengthen the chains that restrain evil. Depending on one's belief system, those women working in closets saved more than an art form.[10]

Pysanky are women's magic. Traditionally, only women and girls made pysanky while working in secret, "lest someone cast an evil spell on the egg," according to New York's Ukrainian Museum. Women decorated eggs in special rooms at night after the children went to bed. The ritual of crafting often began and ended with prayers and was conducted in holy silence. Women carefully selected specimens—fertilized eggs from a chicken laying for the first time. The fertile eggs were key since finished pysanky have a sort of sympathetic magic. Infertile eggs might encourage infertility, and draining the shell of its vital juices before decoration might drain the object's potency.[11]

The decorated eggs represent the springtime return of the sun, fertility, and bounty. Pysanky can be displayed in houses year-round, and plenty of superstition surrounds them. People buried them by doorways to keep away evil spirits and kept them in barns for the protection of animals. Ukraine has a museum dedicated to pysanky and a collection that spans more than ten thousand specimens. As Yaroslava Tkachuk, director of Ukraine's Pysanka Museum wrote, the eggs were once a cure-all. People wore them around their necks as a curative for serious illnesses such as epilepsy, and they rubbed them against their freckles to make them disappear. Young women gave them to eligible bachelors. Each finished egg represented hours of work, and given a young girl's social obligations to give eggs to multiple bachelors, sometimes she couldn't keep up. Many villages had an older woman, someone adept at making pysanky, who sold them to girls in need.[12]

Pysanky are ancient. Traditional pysanky designs resemble the figures on late Stone Age pottery, including geometric patterns and plant and animal life. In Lviv, archaeologists uncovered ceramic pysanky dating to the Trypillian era (5500–2750 BCE). Across Ukraine, researchers have found pysanky on ceramic eggs dating to the ninth century. Surely these symbolic pysanky suggested humans had also decorated real eggs. In 2013, archaeologists discovered the oldest real pysanky on record, a goose egg decorated with a wave pattern that was about five hundred years old.[13]

Of all the oldest motifs, my favorite is the goddess, sometimes called the *berehynia*. Ukraine has an ancient history of empowered women. The *berehynia* motif represents a hearth mother, and historians have linked its presence in modern Ukrainian society to a matriarchal culture existing in the region at around 3000 BCE. As time passed, the figure of the hearth mother became connected to the nymphs who protect riverbanks, the *berehy*, the two melding into the modern *berehynia* around the nineteenth century. According to a paper by feminist scholar Marian J. Rubchak, women in tenth-century Kyivan Rus were "reputed to have viewed this Christian religion [Eastern Orthodoxy] as a misogynist confession, founded on the principle of masculine authority that denigrated them in dogma and right." The women subversively resisted the religion, and perhaps this woman-centric tradition of egg decoration and the *berehynia* design itself is a vestige of that resistance.[14]

The *berehynia* often appears in stylized versions. Hash marks compose her head, and she has two legs, two upturned arms, and two mystery appendages that extend outward from her hips. Artists, ancient and modern, frequently abstracted her into a plantlike form of curls and whorls. But my favorite rendition looks downright pornographic. Picture an oval with pointy ends and six tentacles arrayed, three to a side, around it. Something like a single stalk of wheat is superimposed on the figure—a long straight line from the base of the oval, up to the top, with the wheat head cocked to one side. The overall effect is positively vaginal—a vulva and clitoris. The *berehynia* is said to have power over life, death, water, and fate. In other words, she is mother.

As Christianity became widespread across Eastern Europe, practitioners wrapped new motifs such as churches, crosses, and fish into the pysanky tradition, and reinterpreted older symbols to conform to Christian ideology, a patriarchal meaning written atop a quintessentially feminine art form. Who is the Virgin Mary, with her cockeyed halo, inside a mandorla, but the *berehynia* in deeper resolution?[15]

The famous but, in my view, boring Fabergé eggs layer capitalism on top of the Christian and pagan meanings of pysanky. Between 1885 and 1916, the Romanov czars paid the jeweler to make complex, sparkly versions of pysanky that opened to reveal jewelry and symbols of wealth and narcissism like, say, a golden yolk containing a golden hen wearing a reproduction of the imperial crown or a miniature portrait of the emperor under a diamond, all made of the finest materials. They are expensive simulacrums of the local folk art and beautiful objects of art fit for royalty, no doubt. But rich guys have been doing this forever. Call me when one spends years mastering the actual folk art while also doing his fair share of the child care, and I promise I'll get excited about it.[16]

Although we're not Ukrainian but mostly Germanic American mutts, I wanted to try pysanky with my mom. I secured some tools thanks to the global internet—some serious batik fabric dyes, a small metal funnel strapped to the end of a stick, and a block of beeswax. Mom and I set out to become pysanky experts, hitting the internet for videos. Our favorite featured Miss Justina Marie, a calm, religious, and lovely Ukrainian American woman. She explained that her mother taught her the art, which she practices year-round but especially at Easter. The video included lingering shots of lacquered pysanky—some on chicken eggs, others on larger specimens—in baskets scattered around her house. "Oh, her work is so beautiful," my mom murmured, and I knew I had her hooked.[17]

Miss Justina Marie walked us through the process, from choosing an egg—smooth, uncracked, with no thin parts—to design and execution. She proposed doing this in the traditional way, using whole eggs rather than blowing out the shells. If you store the eggs properly, at a coolish room temperature, in a place where air can circulate, over time the contents will dry up, she said, and if they break, they exude only a minor smell. People do blow the eggs when they've finished decorating them. (If you blow them before dyeing, I had read elsewhere, they turn into little balloons that are hard to sink into the dye.) Blow them at the end, and your masterpiece of hours or days may crack. We tried out our nice, new egg blower on one of the eggs. After huffing and puffing, at last we had an empty shell. Too much work—we'd keep our eggs whole. Our soft-voiced instructor also informed us that modern pysanky uses toxic chemical solutions suitable for dying fabric. Pysanky are for looking, not eating.

Miss Justina Marie explained the basic process. Start with a nice room-temperature egg. Pencil a design onto it. Wax the areas you want to remain the color of the egg. Dip the egg into the first dye, which is often gold. Wax the parts of the egg you want to remain gold. Put it in the second dye, say, light blue. Wax the part you want to remain blue. Continue waxing and dyeing until your design is done. There is an established order of dyes, as some will cover up others, going roughly from light to dark, beginning with gold, moving through pale green and blue, to orange, and then the deep background colors like scarlet, royal blue, or black. The ultimate backdrop is not simply a product of the final dye but of the dyes that came before. My scarlet dye turned a white egg a fiery red, but when it covered yellow or blue colors, it came out more of a brick shade. While pysanky dyes encompass a range of hues, a simple egg might use only one to contrast with the color of the eggshell. Advanced artists have figured out how to complicate the dye sequence with tricks like rinsing off stronger colors in orange dye to move backward in the dye order or dabbing colors over small areas with cotton swabs—far too complex for my mother and me, at least at this stage.[18]

We are not naturals at freehand art. Once, when my dad arrived home from work, I flung my eight-year-old self into his arms and wailed, "Momma is making me do an ART project." My mother has an excellent visual sense: she grows a gorgeous garden, sews quilts, needlepoints, and arranges lovely bouquets of flowers. Over the years we've folded origami, wound embroidery floss around toothpicks to make worry dolls, and done plenty of needlework. You know, women's art or, as it is usually called, craft. But drawing? Not her forte.

Nor is it mine. I am a lifelong asthma sufferer; my childhood medicine made my hands tremor, and I never caught the knack for precise lines. At twelve, I finally figured out how to doodle a passable mad scientist with a triangle head and stopped drawing after that altogether. I love art, though, and even won my high school's art history prize; since then I've been obsessed with art museums and the objects they contain. Now that I have a toddler, I have lots of crayons around, and with his low standards for pictorial achievement, my attitude is slowly changing. It helps that I can draw much better than he can—for now.

The process of waxing and dyeing takes good eyes, steady hands, and an amount of time that ranges from hours to days, as the dye must dry thoroughly between waxings. The trickiest part in our opinion was applying the wax. Artists use a tool called a *kistka*—a tiny metal funnel affixed to the end of a stick—and some beeswax, often dyed black or dark blue for visual ease. The artist holds the funnel close to a candle flame, fills it with bits of beeswax, and then uses the tip of the funnel to draw the melted beeswax across the egg. To maintain the temperature of the beeswax, the writer returns the *kistka*'s tip to the candle and adds wax as needed, perhaps every inch or so. Voilà, a wax pen for writing on eggs. And that is what "pysanky" means; it comes from the Ukrainian verb "to write." I began to feel more hopeful about the project. I don't draw, but I do write.[19]

After writing and dyeing is complete, you've got an egg of jumbled colors, thick with dark wax. The final step, Miss Justina told us, is melting off the coating. You do this slowly, as you don't want to cook the egg in the process. She held her waxed sample egg up to a candle for a brief interval, then buffed off the melted portion with a soft cloth. Bit by bit, her beginner design, an eight-pointed star in white and blue, appeared, which seemed like pure magic.[20]

As soon as my toddler went down for a nap, my mom and I broke out the eggs and pencils, the beeswax and vibrant jars of fabric dyes, and set to work. Perhaps the art is traditionally done in silence, but my mom and I, we're talkers. While she's a private person, she jokes that she's a "talkative recluse." Get her going, and she'll tell you about all kinds of interesting things, like soil composition or the novel she's reading. In our first foray into this ancient art, we decided to copy Miss Justina's "beginner" egg. We began by using a pencil to make guide marks on the egg, sectioning it in half crosswise and lengthwise. Where the marks crossed, we subdivided our quadrants until they became sixteenths. No erasing allowed, as rubber on the eggshell causes dye to adhere inconsistently. Just let the mistake stand and move on, advised Miss Justina. Her expert hands penciled her flawless egg in moments; we spent the better part of an hour smudging pencil into drunken stars on the curved surface. No shortage of expletives escaped our lips.

I felt hopeful about pysanky, though, since it relies on geometric shapes. I can't draw an apple, but a triangle? No problem. Well, almost no problem. As my mom and I discovered, drawing a triangle on paper is quite different from drawing one on a sphere, and that is easier still than drawing one on an elongated egg. We relegated many draft eggs to the "omelet" pile before we felt satisfied enough to move to the next phase.

With high hopes, my mom and I made our clumsy attempts. We practiced waxwork on paper towels, proceeding to eggs as we got comfortable. You must work fast, before the wax hardens. I could tell when my church-lady mom, a twenty-five-year veteran teacher of Lutheran Sunday school, put *kistka* to egg, though, because her first word was "shit." A big splodge of wax marred her pencilwork. Miss Justina had told us how to correct this, by taking a little screwdriver and scraping off the wax along with the top surface of the eggshell, but by now we were well accustomed to living with our mistakes, plus who knew what might happen if we introduced screwdrivers into the process. We upgraded our equipment two days later, from the battered metal funnel included in the twenty-dollar starter kit along with beeswax and six dyes to a pair of better-quality brass *kistkas* purchased for a similar charge from a specialist website. The brass *kistkas* had finer tips than the crude funnel; they also took the heat more evenly and came with a fine wire to dislodge clogs. I also bought two dozen jumbo-sized eggs, as a larger canvas seemed helpful. The new supplies improved our technique considerably, though proficiency still eluded us.

My mom preferred to work in only one or two colors, drawing spirals and stars, and on one egg a very friendly cartoon face in profile. I had no idea she could draw this well, and she informed me that it was her equivalent to my triangle scientist. I tried to copy a historic windmill design, figuring that an all-triangle egg would be easiest, plus the drawing that came with our dye packets showed it in my favorite color combo of red, white, and black. The egg's curvature rendered the triangles tricky; some ended up tiny, others large. After the outline of the windmills went into the first dye, their unfortunate resemblance to swastikas became clear. Perhaps this was more than a resemblance, as swastikas are an old Slavic symbol. I hoped no one would notice. When the egg came out of the dye, my mother cleared her throat and asked me how it was going with my Nazi egg. Fortunately, after additional dye dips and waxed stripes, the resemblance faded.

We worked every day at my son's naptime, waxing new eggs while old ones rested in various dyes. Our initial fumbling had given us a working grasp of the process. We also began to understand why many pysanky makers use a flexible measuring tape to create shapes of the same size, while others with a good eye just . . . eyeball it, which is easier for certain design elements. We started to get the hang of things. I made a sloppy star shape seem intentional by putting a red heart in the middle and adding radiant beams of yellow and red. It looked like some sort of Catholic icon, a haloed beating heart. My mother managed to scribe an excellent poppy onto an egg, but as she lifted it out of the red dye, it slipped and cracked. More expletives. We tried to save the shell at least, using the egg blower. But neither of us had the heart to finish it or throw it away.

The final, wax-melting step took much longer than anticipated and proved monotonous. The busier the decoration, the heavier the wax, the longer the process. My heart egg took upward of twenty minutes to reveal its design. At the end of the week, we had a half dozen eggs, some with simple spirals, others with triangles (and *just* triangles, thank god), stars, wheat, and a jaunty face.

In the weeks to come, the eggs sat in a bowl on my table, reminding me of this time spent with my mother, focused but laughing, as my baby slept. And perhaps that is the magic of pysanky: it fills a bowl with maternal love, fragile, elaborate, effortful maternal love, love I could drink in with my eyes anytime I wanted to, or hold reverently in my hand, silently repeating to myself, "My mother made this. I made this. My mother and I made this, together."

Later that summer, I moved, and the pysanky became submerged in a sea of boxes. As we slowly unpacked, I noticed a putrid aroma near the second-floor bathroom. We isolated the box it seemed to emanate from, a box that unfortunately contained my cardboard carton of pysanky. Many of them had burst, revealing partially cooked egg whites. Perhaps this was our fault for keeping them over the flame too long when melting the wax off. Certainly, keeping them in a hot, enclosed box had been a bad idea. They smelled appalling; like rotting garbage, and the odor clung to the other items in the box and, later, to the porch where we attempted to air them out. A few of the yolks had turned green, and they swarmed with maggots, which probably explained my new house's bottle fly infestation.

Pfarewell, pstinky pysanky.

I still have my mother's love, though, and the egg-shaped memories.

CLOWN EGGS

My favorite egg painting is Dutch artist Pieter Aertsen's *The Egg Dance* because it captures a weird play tradition. A group of peasants gathers in a 1552 brothel for an evening of cavorting. In the foreground, a drunk guy on a stool drapes one arm over a young woman's shoulder. Is he feeling her up? We'll never know since his lifted leg blocks our view of the good stuff. He hoists a tankard high, inciting envy from a bagpiper—whose instrument represents the male anatomy—in the background. The woman seated next to the john seems content to accept his pass. One of her hands rests on a large, empty basket, perhaps a metaphor for her virtue. With her other hand, she tries to direct his attention toward a dancer, who presents a real WTF moment for the modern viewer. The lanky young man hops on one foot amid a floor strewn with leeks and flowers, wooden shoes, a hat, and a sword. Aertsen has captured the youth mid-prance, hands on hips, and balanced on one leg. The dancer does not focus on his delighted audience but on a small chalk circle, an upturned bowl, and an egg resting on the floor.

Getting drunk and trying to move eggs with your feet while dancing passed for fun in the Renaissance. The detritus-strewn floor is an obstacle course, which the egg must not touch. The game's object was to keep the egg intact, roll it into or out of the chalk circle (rules varied), and finally, to place an upturned bowl on it—all with your feet. The winner took home a prize, often a basket of eggs. The Rijksmuseum in the Netherlands houses Aertsen's painting. Its caption adds, "This 'pointless' amusement, along with the dissolute behavior of the other figures, served as a moral warning against debauchery." As with most such warnings, this one is at least half advertisement. Jan Steen's boisterous painting *The Egg Dance: Peasants Merrymaking in an Inn* (1670s) makes the metaphor more explicit. A mass of carousers packs the room. In the foreground, a guy is passed out on an upturned bench next to a toddler and another phallic bagpiper, perhaps dueting with the fiddler behind him. In the background, several women and men have joined hands to dance around an egg. The dancer we can see best is a woman with arms spread and head lolled to one side, an orgiastic smile on her face. She seems like she practically could have laid the egg they are dancing around, a symbol of her discarded virtue. The dynamic crowd around her has certainly discarded their virtue. A couple stumbles upstairs together, while nearer the doorway, two seductions are in progress.[1]

There are other egg dances, of course, among them spring egg dances in which couples minced through a field littered with eggs, as well as stage acts in which a blindfolded performer would dance lightly around a ring of eggs. All of the egg dances are playful, pitting the substantial and sometimes clumsy human body against a fragile shell, much as a springtime egg race or egg toss does. Eggs can certainly be amusing, and to me, nothing captures that quality better than the Clown Egg Registry.

Imagine for a moment that you have contacted the collection's archivist, Mattie the Clown, who will video chat with you from his basement in central London. The basement has plenty of windows and is filled with light and clown paraphernalia. Along one wall rests a large glass case with six shelves and red backing. Each shelf contains a row of tiny plinths covered with dark-colored fabric. Atop each plinth, a clown's face is painted on an egg. Many of them have molded noses, often red ones, affixed. Colorful bits of wig have been glued to some of the shells, along with tiny hats, conical or stovepipe. These are not generic clowns, though: each is labeled with the name of the person it represents. Grimaldi, the father of clowning, has a white-painted face, with bold red triangles on his cheeks and lines drawn around his mouth in suggestion of a chin. His egg has three shocks of red hair glued to its crown. Coco, perhaps Britain's most famous postwar clown, has a bulbous, putty-colored nose affixed to his egg, which features a flesh-toned face accentuated by white-painted eyes and upper lip along with a black U of a mouth. The egg-face sits atop a wine-colored velvet plinth. Lulu Adams, meanwhile, who may represent one of my granny goals, has a white face, wispy white hair, red-rimmed eyes with long lashes, and a conical white hat adorned with three red ribbons perched atop her tiny eggy head.

The Clown Egg Registry arose together with the UK–based group Clowns International and immortalizes famous clowns from history and their membership. It is the world's oldest and largest clown egg collection. According to the clown code, each performer must have unique makeup and costuming. The portraits, captured on real eggs until accidental breakage—particularly of the original collection—motivated a switch to ceramic ones, record these in three dimensions and provide an informal copyright code.

To understand the collection, one must understand the history of Clowns International and the nature of clowning itself, which is why I video-chatted with Mattie the Clown aka Mattie Faint. I list his clown persona first thanks to this story: He used to work at a resort in Barbados as chief public relations officer and events organizer. During his lunch hour, he switched into clown face to run a daily event for children on the pool island. One day, dressed in his office wear, he was returning to the bingo lounge when, as he described, "I pass this mother and her son who had been on the pool island at lunch time, so I said in clown voice, 'ello." "Who's that, mummy?" the boy asked. "Oh, that's Mattie the Clown disguised as a man," his mom said. "I certainly want that on my tombstone," Mattie told me. "Here lies Mattie the Clown, disguised as a man." On the day we spoke, Mattie the Clown was disguised as a kindly British gentleman, with a narrow face and mirth hovering around his eyes and mouth.[2]

The history of modern clowning begins with Joseph Grimaldi, the Regency era pantomime actor who transformed the stock character, Clown, into a stylish performer. He innovated whiteface makeup in his role, which made his features more visible in the dim candlelight that lit the era's stages. He named his character Joey, earning so many accolades for his performances that whiteface clowns are sometimes known as joeys in his honor.

British clowns still commemorate him during a time of year that marked neither his birth nor death. Circuses once crisscrossed Britain during warmer seasons but closed shop in winter. The booking season began in February, when performers converged on a small neighborhood in London to line up contracts. At the first postwar meeting of the clowns in 1947, circus owner Billy Smart had the bright idea to gin up press by asking his clowns to lay a wreath on Grimaldi's grave. At that first gathering, the clowns decided to form a club that grew into Clowns International. And Stan Bult, who painted portraits of clowns on eggs for fun, became the organization's founder and first secretary. From that point on, each February a London church has held a Grimaldi service. The clowns attend in full costume and remember their comrades who died the previous year.

At that same service in 1977, Mattie's first, he made a splash among the clown community. The press's lighting gear blew the fuses in the church hall, and everything went dark. Mattie stepped forward and did a few pocket tricks. "I used to have lights on my bowtie and a red nose that flashes," he said. "So I just stepped forward and stood in the front of the stage and did an act." It felt like hours but probably lasted minutes. After that, he was well known to the clowns. When he moved back to England in the mid-1980s, he joined the group and soon grew involved with the museum, which includes the Clowns International Egg Registry.

By the time I spoke with Mattie, the museum had lost both of its spaces due to flooding. Half the collection had been displayed in the old boiler house in a church basement that flooded. The other half, kept in a cave-based theme park attraction in an old mill, called Wookey Hole, suffered the same when water leached into the museum room's walls thanks to an unknown mill wheel behind them. In 2020 the Clown Egg Registry, like most of us, was unmoored.

Over the course of a few hours, Mattie told me about his life and showed me artifacts from the storied history of clowning. A pig-pile of oversized shoes, some of them dating to the 1930s, sat atop a wardrobe. A rack of costumes included an elaborate whiteface outfit decorated with rickrack that dated to the 1860s. We examined Coco the Clown's last costume. One pocket contained a lottery ticket from after his death, when his son, who designed the makeup and outfit for Ronald McDonald, had worn his dad's costume. As Mattie told me, "Whiteface is very circus" and thus rather out of date, especially since you can't use it at a children's party: the costumes have no pockets and are too valuable to get messy in.

There are a few basic types of clowns: whiteface, Auguste, and character. The whiteface clown—essentially created by Grimaldi—is the oldest of the three and comes out of commedia dell'arte traditions in which stock characters such as Harlequin, Pierrot, and Clown perform. They tend to be elegantly dressed—their costumes can cost a fortune—and they often served a central role in organizing troupes of clowns on- or off-stage. They are primarily seen in the circus. As their name suggests, their makeup features a white background on which other features, traditionally including exaggerated and asymmetrical eyebrows, are painted. Their job is to set up the jokes for the Auguste, but they tend to stay clear of anything foolish or messy, like water or wallpaper paste, which might make their fabulous costumes suffer.

Auguste clowns are buffoons who paint exaggerated features over flesh-toned makeup. They've got red noses, wild hair, loud and large outfits. They take cream pies to the face with astonishing frequency. When most people think "clown," they are probably picturing an Auguste. It's likely that the Auguste was originally intended to be a drunk. According to *The Clown Egg Register,* a photographic book by Luke Stephenson and Helen Champion, one origin story for the name dates to the 1860s. At Germany's Circus Renz, an injured American horseback rider, Tom Belling, put on the circus owner's costume backstage and "began mimicking him. The owner caught Belling in the act and an angry chase began, ending in the circus ring. Believing this was part of the show, the audience began to shout *Auguste!* which is German for 'fool.'" Mattie tells me that people watching Augustes these days might not realize it, as the heavy face makeup has fallen out of favor. Modern Auguste clowning includes characters like Mr. Bean, who has "beautiful stylized movement," Mattie said.[3]

Character clowns have professions, such as doctor clown, police officer clown, or firefighter clown. Their makeup and dress vary but make their professions clear. The Hobo or Tramp clown—picture Charlie Chaplin—is sometimes described as its own clown type but could be considered a down-on-their-luck character clown as well.[4]

Mattie's clown is an Auguste, eyes of simple black crosses on a white field, with a white muzzle, demure red mouth, and enormous red nose. A jaunty sunflower sticks out of his bowler hat, and he wears a yellow-and-red plaid suit and a large, red bowtie speckled with white polka dots. He has spent decades clowning at birthday parties and as a clown doctor for children in hospitals.

Finally, laid carefully against the wall in a clear case, Mattie shows me much of the Clown Egg Registry and tells me about some of its artists. "Stan Bult . . . I reckon he was painting eggs before the war, because he was a clown and circus fanatic—the war was just an interruption." Bult became the leader of the Circus Clowns Club at the first postwar meeting of the clowns. After he died in 1966, Jack Goff stepped in to paint eggs "just to keep the system going, and then he died," Mattie said. Many of the older eggs have been broken or lost over the years. Mattie first moved to London in 1969, and while he wasn't a clown yet, he was interested in clowning, so he visited a restaurant in the West End called Clown. "And I remember seeing a lot of eggs in a show case, and then the restaurant closed, and they disappeared." Decades later, the granddaughter of the restaurant manager found some of them, including original Stan Bult eggs, and sold them back to Clowns International. In the mid-1980s, the organization revived the tradition, and artist Janet Webb began painting eggs, completing around forty before the role passed to Kate Stone and, from her, on to Debbie Smith.[5]

Before I met Debbie, I had already fallen in love with the playful nature of the clown eggs. I've been a student of play all my life, from a childhood spent enticing my parents to play cat family with me to the subject of my first book, the storytelling medium of larp, a sort of folk theater in which participants portray characters going through an improvised story arc. Like clowning, larp is an ephemeral art form, an experience that happens in real time (albeit without an audience). As doctor, psychologist, and play researcher Stuart Brown writes, "The opposite of play is not work—it is depression." For me, play is life. It is the wiggle and jiggle room on a steering wheel—the moment when you haven't yet decided to go right or left and so everything is permitted; it is actors onstage being most themselves they will ever be while dressed in the skin of another. Play, like the egg (both alive and not-alive, precious and worthless), holds a superposition of states. It can be quite serious—witness an athlete focusing before a game—but at the same time, although we act seriously, play is fun, full of pretending, sealed off from the outside world by what scholars call "the magic circle," the physical and emotional bounds of a game that separate it from the outside world. If you lose at chess, no one dies, unless you are unwittingly the character on a TV mystery. The magic circle is the board of chess, the proscenium of a stage, and the psychological space created by the special rules for behavior in such places. Like an egg, the magic circle is not hermetically sealed: gas passes to and from the developing chick, and the emotions around play tend to bleed out into reality. We cry when Romeo drinks the poison, friendships can end over a game of Monopoly, and when we exit the movie theater, our lives may be irrevocably changed by what we've seen.[6]

Scholars have argued that play is something we do for its own sake. By this metric, the clown eggs certainly fit the definition.

Ostensibly a copyright tool, the registry has no truly practical purpose beyond historical record, especially given what Debbie told me about clown makeup. It's "based on faces and the lines of your face," she said, so even if you copied another clown's makeup, it would look different on you because your face is different. The eggs, on the other hand, are basically the same— smooth canvases to showcase that makeup.[7]

And since I'm in love with the idea of clown eggs, the fragile shells that fix the ephemeral art form of famously clumsy clowns, it is only natural that I should develop a small crush on the artist herself and the story of how she became the official artist of the Clown Egg Registry.

An unassuming British woman with medium brown hair and bangs, Debbie has mischievousness in her eyes and around her mouth, which I suspect might be a clown's hallmark. She began as a fine artist specializing in portraiture and used restrained palates of subtle color in her work, which she exhibited in local galleries. But on a fateful Sunday in 1989, the newspaper arrived with a full-color supplement inside. She opened it, she said, to find "a double spread of a clown convention, which I'd never heard of. . . . There were hundreds of clowns in this one big picture." She'd always kind of liked clowns, and as she examined the picture, she began to think about her color choices and how fun it would be to use the bold colors of the sharp bright makeup.[8]

From there, the slide into the clown world was swift.

More than forty years after its origins, Debbie visited the Grimaldi service. She met a clown there who suggested that if she wanted to paint fools, she really should come along to a convention. She agreed and scored a last-minute place to display her work after another vendor bowed out. That display earned her a few clients. Soon she became a clown scenester. Whether domestically or abroad, "For a few years, I used to go around to all the things clowns were at," she told me. And, of course, if one finds oneself at a clown convention already, one might as well sit in on a clowning workshop or two—you know, only to understand your subjects a bit better. And if one is attending clowning workshops, well, it's only a matter of time until one discovers her own inner clown and begins to workshop her own makeup. Jolly Dizzy shares a bone structure remarkably like Debbie's but has a white muzzle and eyes and a red heart on her nose. She is available for corporate events as well as children's parties. And if one begins as a fine artist and becomes a clown, one might as well marry another clown, which is what Debbie did. And then, in 2009, Kate Stone retired as the official artist for the Clowns International Egg Registry, and Debbie threw her cream pie into the ring. The rest, as they say, is history.

Debbie made the short slide from artist-spectator to clown to clown artist almost in freefall. Clowns have a PR problem: they are maligned in popular media, unjustly so, according to participants. As Debbie explained to me, "There are so many different types of clowns, but all clowns, their aim is to make that child or person laugh, and it doesn't have to be an outward laugh. I've done a lot of clowning with special needs children in hospital," where the amusement might just be "a glint in the eye." The essence of clowning is laughter, and the media focus on scary clowns, well, "It's so annoying," Debbie said.

As for the work of making the clown eggs, the portraits can take anywhere from two days, which is average, to a couple of weeks if, as Debbie explained, "you want to make a little tie to go with it, or just get carried away with doing the detail." Debbie paints two copies at once—one for the official collection, presently housed in Mattie's basement, and one for the clown to have. She begins with photographs of her subject—full face and profile—along with swatches from their costumes and wigs. Often, she does a little bit of a sketch in pencil on the egg first. How she approaches the egg depends on how the clown approaches their art.

For Debbie, painting the exaggerated makeup of the white-face clowns is easier than rendering an Auguste or character clown with more minimal makeup. Those are more like painting a portrait. Still, turning a two-dimensional picture into a three-dimensional egg isn't easy to do. That's why Debbie asks for full-on and side-view portraits to help her, and sometimes ends up sculpting features like noses. It's best if she can paint the eggs as the clown poses in front of her, but that is not always possible. She also asks her subjects questions. Eye color, for example, doesn't show up well in photographs, but she needs it for the egg. And then she painstakingly re-creates the costume on both the egg and, breaking with previous tradition, the plinth. The costume-making requires its own special skill. As Debbie told me, "Because I was doing the clown eggs, I was learning to do miniature hats to go on the eggs." So like a good researcher committed perhaps to overkill, she took three or four millinery courses. "You have to make little molds for the miniature bowler hats; I have to make them myself and stretch them."[9]

Over the years, the collection has had five artists (Stan Bult, Jack Goff, Janet Webb, Kate Stone, and now Debbie Smith), each with their own style. "When I started doing mine, some people said my eggs must be bigger than others," Debbie said. It wasn't until she saw her eggs next to those of the previous artist, Kate Stone, that she realized her podiums were taller, and that while Kate made miniature hats perched atop the egg, she had been making hats the same size as the egg, so the scale seemed larger, although the eggs weren't.

The Clown Egg Registry may sound playful, but it is also serious, Mattie reminds me. The registry is the stuff of law school

faculty dreams; for the *Notre Dame Law Review* two professors once authored a sixty-nine-page paper on it as an example
of informal copyright codes. The paper contains entertaining
subheadings like "Clown Norms," interspersed with legalese
like "Property Registers Beyond Exclusion." As far as I can tell,
a clown egg has never been used in court as part of any copyright dispute around makeup or costuming. But it is a historical
record of clowning, however unconventional. As Debbie told me
over the phone, clown faces belong on eggs because clowning
"is an art, so why not record it in an artistic way?" The clown
eggs take an ephemeral discipline and preserve it for posterity;
there is playful irony in doing so on a fragile eggshell. As Debbie explained, "You think about clowns being fragile as well,
although the eggs we do now aren't fragile. But the idea of an
egg being fresh and fragile kind of lends itself," she said. Plus
"the comical thing is that it could fall to pieces."[10]

Over the years, the collection has been shown in exhibitions in
Bristol and the Netherlands and immortalized in *Men at Arms*,
a Discworld novel of the late author Terry Pratchett. Members
of Clowns International can apply to have a ceramic egg painted
with their portrait for a very reasonable fee. Yet Debbie has
painted only twenty-eight portraits in her dozen years as official
artist, as the popularity of clowning and live entertainment in
general, she said, has waned. Still, she has always admired the
artists who did this work and loves her place in the pantheon of
chosen painters. She has yet to do her own portrait, though perhaps she'll get around to it one of these days.[11]

It's delightful to live in a world that includes the Clown Egg
Registry. On one level, it represents eggs—tricksters who are
alive and not-alive—perfectly. But the clown eggs are, of course,
a double trickster since the collection is both egg and not-egg
(ceramic). The concept of a copyright registry plays against the

inherent whimsicality of clowns: ironic but appropriate that the clowns have chosen to immortalize their transitory art form in fragile materials. The egg's the place for them, even though—or perhaps especially because—eggs break as indeed much of the collection has. The clown eggs are something of a memento mori, a reminder that we carry the seeds of our own destruction within us. We must be gentle with the clown egg collection, with the art form of clowning, an art form that is perhaps dying—no, not dying, Mattie says when I read him that line, but changing to fit the modern world. I am reminded of the tarot card for death, which I am told may not mean literal death but a metaphorical shift. The clown eggs tell us that no immortal refuge exists, not even in art. Every omelet leaves scattered shells in its wake. The white of the shells represents, if not purity, then maybe the pale color of a skull. Alas, poor Yorick, we knew him, the king's jester, and we'd have known him better if his clown egg had survived. But it didn't. And neither did Yorick. And neither will we. Might as well slap on some pancake makeup and play the clown.

EGG TOSS

In an art gallery in Beijing or Berlin, Los Angeles or Mexico City, an artist stands before a specially treated white wall. She has jaw-length graying hair, sharp eyes, her body concealed beneath a large sweatshirt. Before her stands an assembled crowd and a thousand eggs. She gravely instructs them on how to be mindful with their missiles.

In a brutalist plaza before a German art museum, or on a seaside cliff in Mauritius, or inside her art studio, another artist stands, naked. She has a blunt, raven bob and the idealized body of a professional model. She straddles her canvas and begins to squat.

There is just something about eggs and conceptual art. Perhaps it is their utter banality: they are as common as a soup can or paint pot. Perhaps it is their fragility, wrapped in an architecturally sound dome. Perhaps it is their connection with femininity and its regenerative power. Or perhaps it is lust for a disrespectful mess.

People have tossed eggs for centuries. Throwing eggs is fun: witness the popularity of Easter egg tosses and Halloween vandalism. Since Roman times, Team Egg and Team Flour battle for dominance each December in the Spanish town of Ibi, which ends in a glorious mess. Renaissance eaters used to clear the air after dinner by hurling eggs filled with perfume. A tossed egg is also political, a sign of supreme disrespect. In medieval times, passersby hurled eggs at those imprisoned in the stocks, while in 2020, more than two thousand people lined up to egg a new statue of Margaret Thatcher. Australia's "Egg Boy" earned viral fame in 2019 after yolking a right-wing politician who made a racist statement after a mass shooting at a mosque. The boy threw eggs; the politician punched him in the face.[1]

Perhaps my favorite political egging is that of Françoise Dior, niece of the famous fashion designer who rejected her due to her politics. A wealthy, prominent French neo-Nazi, she visited British neo-Nazi groups and struck up friendships—and romances—with their leaders. She wed one of them, Colin Jordan, at a registry office in Coventry in 1963. In a later ceremony at a private residence festooned with Nazi regalia, they cut their ring fingers and let their commingled blood drip down onto a first edition of *Mein Kampf*. After both ceremonies, the newlyweds stepped outside to face a crowd of protesters, who hurled rotten eggs and apples at them while Jordan and Dior gave the Nazi salute. The right brings guns; the left brings eggs and weaponizes mockery.[2]

All of these—the fun of eggs, the disrespect of a tossed egg, and the egg's rich symbolism—make it a powerful medium for art. Before we delve into tossed egg art, understand that eggs have long been fundamental to the artistic process. For centuries, they have been used in paints and varnishes. In the kitchen, egg yolk binds fat and water into a creamy colloid; eat a fried egg on a plate, and you'll find it also binds itself to crockery quite well (dishwashing hack: use cold water so the heat doesn't cook it onto the surface). The same concept allows eggs to bind pigment to board or canvas. Simply mix a yolk with a colorant and a thinner, usually water, and you've got a fast-drying paint that

can stand the test of time. Although it has a few older anteced-
ents, egg tempera paint dates as far back as the 1200s. Giotto,
widely regarded as the father of European painting and the first
artist of the Renaissance, painted using tempera, which enabled
his famous advancements. He was one of the first to use perspective—foreshortened limbs—and naturalistic faces in his portraits. Over his career he painted many gilded icons of Madonna
and child using gold dust mixed into shining yolks.[3]

The white and shell also have their uses, dating back at least
to the fifteenth century. Finely ground eggshells can extend paint
and lend it texture. As for the white, well, anyone who has had
a baking mishap can tell you that egg white, dried on a counter-
top, is shiny and hard to clean up. Artists used egg whites to give
paintings a glossy sheen. Mixed with pigments like saffron and
then dried, they also made cake paints. In the mid-nineteenth
century, meanwhile, early photographers frothed it with salt to
make glossy prints.[4]

But my favorite artistic use for eggs lies in conceptual pieces
that are as much event and idea as object. Famed surrealist
painter Salvador Dalí, best known for his floppy watches, had
something of an egg obsession. He painted melty fried eggs,
fried eggs hanging from strings, boiled eggs with faces, and
celestial eggs sprouting flowers or waterfalls. The museum/tomb
he designed for himself features huge concrete eggs in parapets
and jumbled across his tower. The building even contains an
airy, glass-domed atrium in the gigantic shape of an egg. Perhaps lesser known are his 1938 plans to construct a ten-foot-tall
boiled egg out of eggs, a project never realized in his lifetime.
Fortunately, culinary historian Carolin C. Young, author of
*Apples of Gold in Settings of Silver: Stories of Dinner as a Work
of Art*, and her friend Charles Foster-Hall, an artist, took up
the project in the mid-2000s. As she writes in a paper on their

efforts to realize this dream, they planned to make "the gigantic boiled egg, from the separated whites and yolks of more than a thousand real eggs, poured into aluminum molds for boiling. . . . Standing in the gallery, it was to be tasted by visitors using spoons three feet long, so that they could dig in as far as the yolk and ascertain that the egg was REAL." The presence of other large eggs made of eggs—but none so titanic as Dalí's—in culinary history, as Young puts it, "reveal that it [Dalí's vision], with Surrealist bravura, masterfully taps into an ancient, if esoteric, human longing for a massive egg." Later, she added, "Why build a giant egg? Why not build a giant egg!" To egg is human, but perhaps a massive egg touches on the divine.[5]

A ten-foot-tall boiled egg is a simple organic concept, but as with many simple designs, its enactment would require spectacular feats of logistics. The volume of the finished product demands approximately 155,000 large chicken eggs and a battalion of gloved volunteers hand-separating yolk from white in a chilled environment. It also requires an enormous egg-white mold in two halves, sprayed with calcium carbonate to create a shell, as well as a two-part spherical mold for the yolk. To ensure even cooking, heated iron rods would have to be inserted into the egg at carefully staggered intervals so as not to undermine the structural integrity of the finished mass of protein. Gravity would keep the egg halves together, with additional calcium carbonate sprayed on the seam and sanded to create the appearance of an enormous, unbroken shell that could be cracked in front of an audience. Displaying the egg would carry its own challenges, requiring an enormous, architecturally sound egg cup. To maintain sanitation, the oversized spoons would rest in a gargantuan glass of sudsy water. Afterward, of course, there would be the problem of disposing of 18,387 pounds of stanky, half-eaten, week-old egg in an ecologically responsible manner.

It's easy to understand why Young and Foster-Hall enlisted the advice of a cadre of experts, including molecular gastronomy guru Hervé This, engineers from the MIT Media Lab, and an architectural team.

Although the dream was unfortunately never realized—the creators ended up with differing opinions about the project's sustainability and wastefulness—I find it tremendously appealing, both because of its ballsy vision and its herculean logistics. I imagine myself at this event that didn't happen, bonding with fellow enthusiasts in the fridge rooms, separating a hundred thousand eggs together. And what a marvelous story to tell my grandchildren. The idea of a titanic edible egg conjures a dream-world banquet without limits. In the mythology of my childhood, a boiled egg really is about ten feet tall.

When I consider Dalí's giant egg, though, I also view it through a gendered lens. Here's another man obsessed with the female reproductive cell, mediated through the humble boiled egg, a crosscultural comfort food most frequently cooked by mom and grandma. It reminds me a little of French artist Abraham Poincheval's 2017 performance at the Palais de Tokyo Contemporary Art Museum in Paris, called *Egg*, in which he played a hen, quite literally. He sat atop ten fertilized chicken eggs for twenty-three days. Nine of them hatched. Before becoming a hen, he spent a week living inside an egg-shaped rock to get into the idea. What a longing for the feminine, to take on the nurturing roles society typically assigns to women—homemaker, nester, nourisher. It taps into desires that the men in my life— friends, family—have expressed to me. They are hungry for representations of masculinity beyond the strong-silent-violent stereotypes. They want to see men like them, who cook nourishing meals, hug their children, and discuss their feelings with other men.[6]

The women, and some of the nonbinary people in my life, on the other hand, want to reject an entirely different social script. They want to be taken seriously and freed from the expectation of 24/7 nurturance. As Jenny Offill's oft-quoted lines from her novel *Dept. of Speculation* say, "My plan was never to get married. I was going to be an art monster instead." Or as my friend—a user experience expert at Google with the weight of her immigrant parents' expectations on her shoulders—put it to me during one of our chats, "I want to move to the mountains and become an art witch." Me too, friend, me too. I aspire to avoid modes of thinking that treat my energy like a zero-sum game. I want to have a fulfilling relationship with my spouse and child and *also* be an art witch. And yet, as I send my toddler away to the tender loving care of his grandpa each morning to give me time to write, I often feel monstrous. That should be me, wiping his sticky little hands and calming his infantile tantrums. And I hate the part of myself that still feels guilty and beholden to this social script.

That is part of why I find Milo Moiré fascinating. She's an art monster par excellence who takes expectations about good-girl femininity and literally rides around town (Düsseldorf) with them, naked, on a bicycle with a red dildo instead of a seat. Her work limns and crosses the boundaries between pornography and art. She's Swiss, of Spanish-Slovakian descent, but lives and works in Germany, doing mostly conceptual art that features her naked, sexualized body. She's raven-haired, leggy, slender, and full-breasted. Over the years she has been arrested at the Eiffel Tower for taking nude selfies in protest of sexual violence and barred from Art Basel while performing *The Script System No. 2*. Instead of getting dressed, she simply wrote the words for clothing items—shirt, panties, etc.—on her body. Her pay website offers a trove of deliberate spank material that her partner, art photographer Peter Palm, has taken of her.

But it is her PlopEgg performances that interest me most. After clicking through a "this may not be for all audiences" warning on YouTube, I found footage of her first PlopEgg staging at Art Cologne 2014. She stands naked on a raised platform. A small, white sheet on a frame is wheeled in to surround her waist. She bends down, focused for a moment, and then a clothed assistant wheels the frame away. YouTube has put black bars over her breasts and vulva. She moves into position, feet planted on two different risers that arch above the canvas, and she begins to squat. She moves her hands into prayer pose and holds them to her face as her body tenses. There is a pregnant pause followed by a crack and a red splat as she expels a paint-filled eggshell from her vagina. The white curtain returns, she inserts another egg and moves into a new position over the canvas, and—crack! splat!—a beige stream appears. She works methodically, choosing positions with care, and sending eggs filled with browns, pinks, and creams onto the canvas. Her assistants fold it in half, and she glides a squeegee over it, breaking any remaining shells. The result is a mirror-imaged, vibrantly colored Rorschach blot. To me, it resembles a red pelvis bone that is menstruating, with mud splatters or perhaps divine radiance emanating out of it; it also looks a little like a bleeding heart.[7]

Moiré is focused, alone, and vulnerably naked in a sweeping concrete plaza in front of an art museum. The spectators are cozied up in fall sweaters and jeans. The scene reminds me of my experience of childbirth, naked and vulnerable in a room of clothed assistants intensely focused on me. Despite the crowded room, I felt lonely: they were here to support me, but I was the only one who could pass this child's head through my pelvis, a proposition that terrified me. Don't scream, said the nurse. You'll lose the abdominal pressure you need to push the baby out. I knew what the baby's head would do to me, and that fact terrified me. But on the second push, I didn't scream.

Moiré also labored. She had made elaborate preparations: obtaining, blowing, and filling eggs; selecting the paint; arranging for the raised planks; preparing the canvas; and training her assistants. Having a chicken egg up her vagina couldn't have been comfortable.

The PlopEgg series speaks to my desire for play and for disrupting "nice girl" social scripts. Ironically, I first encountered the piece through *Guardian* writer Jonathan Jones's epic pan of the performance, headlined "The artist who lays eggs with her vagina—or why performance art is so silly." He calls the PlopEgg performance "absurd, gratuitous, trite and desperate. Anywhere but an art gathering, this would be regarded as satire on modern cultural emptiness." But what's wrong with a little absurdity? Where is Jones's sense of fun? Renoir famously said he painted with his prick, metaphorically at least, so why shouldn't Moiré paint with her vagina?[8]

If Moiré fulfills my art witch longings—she's the kind of sexy, bad-girl performance artist I would have loved to be in my mid-twenties—then Sarah Lucas fulfills a more mature version of my personal art monster fantasy, and one that channels gendered rage.

Many of my women friends feel deep-seated anger because we have been brought up to be caregivers. While caregiving is necessary and often rewarding, it can also be a huge burden. I am put in mind of Cal Newport's otherwise excellent book *Deep Work: Rules for Focused Success in a Distracted World* about the creative power of long periods of solitude. Nearly every example he gives, though, is about a man. In one he mentions Carl Jung decamping to Bollingen Tower—his purpose-built workhouse—and spending months there while his wife solo-parented their five children. Emma Jung was an heiress who employed servants, so I don't imagine she went to toddler hell, which is where the phrase "solo-parenting their five children" makes my mind go. She was also an important psychoanalyst who spent thirty years writing a book on the Holy Grail myth but didn't complete it before her death from cancer. Maybe she needed a tower of her own too.

I and most of my women friends tend to take on the role that one of them has dubbed "Relationship Captain." A relationship captain is quartermaster, cruise director, and general contractor. Feminist theorists call this labor the "mental load," the burden of planning and thinking through logistics: for example, remembering to buy birthday gifts two weeks before the big day so they arrive in time to be wrapped and mailed. Like a lot of so-called "women's work," these duties don't obey standard working hours, and like the dishes or a newly polished kitchen floor, this work is constantly undone. I want to be free to focus on my creative work; I am not. I am endlessly picking up socks and closing cabinet doors and trying to smooth my own way through the yearly birthday gauntlet. It's enough to make me want to throw an egg. And that is why I have a deep love for artist Sarah Lucas.[9]

Lucas came to fame in the mid-1990s as part of the loose collective once dubbed the Young British Artists, or YBAs, a group that included shock jock artist Damien Hurst, among others. Lucas makes work full of visual puns. If Moiré is the sexy art witch I wanted to be in my twenties, Lucas is the art witch I need right now—more of a jester, someone with a storied career who has matured with age without losing her sense of play. Lucas is famous for her sculptures and photographs, including many featuring fresh food. She has made too many bad-ass pieces to cover them all here, so suffice to say that her first show had the title *Penis Nailed to a Board* and that over the years she's attached raw chickens to granny panties and exhibited plaster casts of her friends from the waist down, smoking cigarettes from their navels, anuses, and vulvas, displayed against an egg-yolk-yellow background.[10]

Eggs are a continuing theme and obsession in her work. Perhaps her best-known piece is the 1996 photograph *Self Portrait with Fried Eggs*, in which she manspreads on an armchair with two fried eggs atop her breasts. The visual pun, as the critics call it—fried eggs, slang for small breasts, which Lucas has—is funny enough. But her defiant gaze, which dares you to say shit about it, makes the photograph both silly and profoundly serious. Before *Self Portrait with Fried Eggs*, there was *Two Fried Eggs and Kebab*, a scuffed-up wooden table with the objects arranged to form two boobs and a vulva. When the piece traveled to Los Angeles's Hammer Museum in 2019 as part of her *Au Naturel* retrospective, the museum staff were tasked with frying fresh eggs daily. As the *Los Angeles Times* reported, "The artist gave Pugh [the museum's art preparator] very specific instructions: sturdy-bottomed eggs with browned edges; no seasoning or oil, lest the table they sit on become soiled. 'And they are to be sunny side up,' Pugh adds. 'A little more like a cartoon, more of a hyper-real kind of an egg.'" A real egg playing the role of a cartoon egg in an art gallery.[11]

Finally, Lucas has made numerous works that involve shattering raw eggs. In the film *Egg Massage*, made with her partner, artist Julian Simmons, she attends a lovely New Year's Eve dinner party with friends. Afterward, Simmons lies naked and face down on the cleared table. People bring lit candles, a cut pineapple, and an apple to be part of the still life. On the counter in the background, a vise holds the hoof of an expensive-looking leg of cured ham. The music gestures toward the

high-class. Lucas plays a half-sized violin, and one shot captures a visual duet with a woman carving delicate slices of
ham, knife and bow working in concert. Lucas takes out two
flats of eggs, about fifty of them, and places them in a stack
next to Simmons. She opens their shells, one by one, by rap-
ping them smartly with a knife. Simmons writhes sensuously
on the table as raw eggs slide off his body. Lucas laughs and
comments that there are not nearly enough. Soon eggs drip
down the crack of his butt, and Lucas begins to massage the
unbroken yolks in, with workman-like strokes. The title of the
piece includes "massage," but she looks like she is varnishing
him with egg. The fiddle music continues, as does the laughter
of the party. Soon the table is slick with eggs: Simmons's body
slides around on it as Lucas flips him onto his back. Lucas continues to varnish him, even manipulating his testicles so they
resemble unbroken eggshells. At the end, Simmons poses with
a cardboard flat, filled with broken eggs, over his groin. When
he gets up to go, the camera zooms in on the destruction—
pools of egg and broken shells amid the candles. A dustpan
begins to scrape the egg off the table, and the piece is over. I'd
agree with critic Zachary Small about the overall vibe, which
is "ritualistic and downright witchy."[12]

I found *Egg Massage* to be a rich text. Simmons's writhing
seems familiar—it is unmistakably erotic posing—although I'm
unused to seeing a man in a position like that, splayed out for
the viewer's pleasure, objectified, and perhaps a bit degraded.
Lucas covered him in eggs with the disinterested precision of an
artist. She laughs and pays attention to where the yolk is on Simmons's body. But this isn't merely light fun. Another critic, Digby
Warde-Aldam, explains, "Everyone caught on camera seems to
be having the time of their life, yet it's impossible to watch without feeling a little bit uncomfortable." As with much of Lucas's

work, I respond strongly to her sense of play. She turned the female nude in art—a woman reclining with a suggestive gaze—on its head. Instead of the odalisque, the male partner; the violence of the sexualized gaze replaced by a woman artist who has him egged and has got him by the balls.[13]

But my favorite work by far is the cathartic venting of feminine rage on display in the participatory piece *One Thousand Eggs: For Women.* It is what it sounds like—a specially treated gallery wall, and a thousand eggs for women, or people of any gender wearing stereotypically feminine clothing, to pelt at it. The piece has been produced several times in different cities but always uses the same format. Lucas hands out the first few eggs to spectators and instructs them on how to toss. As Lorissa Rinehart wrote of Los Angeles's Hammer Museum run in *Hyperallergic,* the event began with Lucas addressing the crowd, asking them to cover the wall evenly, to let everyone have a turn, and not to hog all the fun by throwing too many eggs in a row. "For a moment it all seemed like a high-minded exercise," Rinehart wrote. "Then the egg-throwing began." The dull thud of eggs cracking on the wall formed an erratic drumbeat to the proceedings. The air was one of "gleeful destruction" and feminine rage. The piece's satisfaction, Reinhart writes, comes from the participants' emotion: "You're not supposed to throw eggs, women are not supposed to express anger, and we sure as hell aren't supposed to make a mess. Yet by the end, that's exactly what we had done."[14]

There is something poetic in the image of women hurling the world's most precious cell at a solid wall. The eggs are what we are supposed to take care of, what we are supposed to nurture—crack!—but here destroy. The eggs in our bodies are limited, finite, unlike, say, the sperm that testicles keep making—

splat!* This presumption of feminine caregiving underpins laws designed to control the female body and its access to services like abortion, which Lucas had after a teen pregnancy. Lucas created an alternative reality in which women decide what to do with eggs—and use them to make art.[15]

At an artists' colony once, while I was deciding whether to have a child, I met a coterie of female novelists and poets, mothers, who told me what it was like. Each child is one less book I wrote, one said. But I love to see the poet read, and I love to see the actor act, she added.

Every egg used is one less to throw at the wall and see what sticks. But each, perhaps a work of art, has agency in its own right.

---

* With the caveat that recent studies provide a growing body of evidence that human ovaries may be able to generate new eggs throughout the course of a person's life, courtesy of recently discovered ovarian stem cells.

10

SPACE EGGS

Humans are curious creatures. Even small humans are famous for their "why why why" stage, which my own child passed through as I interviewed space experts. My son had not yet realized that not everything is knowable and that of the knowable, not everything is useful or interesting. But we humans still can't help asking questions, especially about nature and the fragility of existence.

Human life has one major vulnerability: our eggs reside on a rapidly warming planet. Dorothy Parker reportedly quipped after her legal hospital abortion that it served her right for putting all her eggs in one bastard. Space travel offers several solutions to the problem, though: it satisfies humanity's curiosity about our surroundings and creates scientific knowledge that can benefit us here on Earth. Ultimately, the goal is to back up the processes of life on another planet, say, Mars.[1]

Living things, of course, need a continuous supply of food, including protein. Any space colony—everything sounds cooler when prefaced with "space," no?—could do with a self-renewing source of protein like, say, a bird that lays edible eggs. As a bonus, since eggs double as proxies for the human body, studies of space eggs function as experiments in how microgravity might impact human reproduction.

The Soviets chose Japanese quails for early experiments. Japanese quails convert feed into eggs particularly efficiently. Only a third the size of chicken eggs, quail ova have greater nutritional density. Gram for gram they provide slightly more protein, fat, and calories; double the riboflavin and iron; and one-and-a-half times the vitamin B_{12}. The Soviets blasted fertilized quail eggs into space in incubators several times. In 1979, the egg embryos aboard the *Soyuz 32* developed more slowly than those on Earth. They were also missing a crucial body part—the head. The scientists thought that their equipment had malfunctioned. Later that year, sixty more quail eggs went up on the *Bion 5/Kosmos 1129* mission. These eggs didn't fare well on reentry to the atmosphere, as the humidifier malfunctioned. The embryos ended up dehydrated but seemed to have developed normally. In 1990 quail eggs aboard the space station *Mir* hatched, the first vertebrates to be born outside Earth's atmosphere. Although it seemed like a success at first, in the absence of gravity, the chicks weren't precocial, that is, they couldn't run around and feed themselves. Instead, they slowly starved to death despite the astronauts' best efforts to hand-feed them. Researchers eventually developed a harness to help the chicks get food, but there were still problems: the birds weren't romantically inclined, a nonstarter as a space colony protein source.[2]

The United States has also sent eggs to space. In the beginning of space exploration, or at least close to it—before *Soyuz*, space walks, moon landings, or even rockets, in fact before we thought about whether we could have ongoing food sources—we had to answer the question, "Assuming we get up to the vacuum of space, will anything there kill us?" In the 1910s, Austrian American physicist Dr. Victor Hess risked life and limb in pursuit of this question. He grabbed an electroscope, jumped into a primitive high-altitude balloon, and floated into the stratosphere to investigate. He inferred that little subatomic particles must be whizzing past, their effects a function of altitude. As his Nobel citation put it, "Surprisingly, he found that ionization first decreased, but then increased again at higher altitudes. He concluded that the upper atmosphere is ionized by radiation from space. He proved that this radiation is not solar through experiments performed at night and during eclipses." Amazingly, he had discovered cosmic rays. As of 2021, it was still not completely clear where much of this space radiation originated.[3]

Fears about cosmic rays sparked many high-altitude balloon experiments, in which scientists sent up countless biological samples—plants, animals, tissue cultures, and of course, eggs—to see if the cosmic rays would harm them. When I began investigating eggs in space, my game design collaborator Jason, a space fan, suggested I join the Space Hipsters Facebook group. Alas, I am only a space poseur. I find space terribly big and boring since it is full of nonliving things like rocks and gravitational waves. But the space hipster group is where I met Dr. Jordan Bimm, a space historian and Guggenheim fellow who kindly guided me to many space egg details. He called the high-altitude balloon experiments of the 1950s "obscure military projects of dubious utility." Basically, they showed that cosmic rays don't cause instant death. And yet, as Bimm wrote, "Space is such a mirror for our culture, when you look at the 'space version' of a thing you really see a weird, amplified, version of it." With that in mind, let's take a quick survey of what went up in the balloons.

Although plenty of items—animals, plants and their seeds (eggs of a different kind), vegetables, viruses, and bacteria—went up, I'm most interested in two items: animal eggs and cultures from human tissue. The eggs included fertilized chicken eggs as well as ova from other animals. As Bimm wrote me, "Chicken eggs were among the first biological objects used to study the effects of the space environment. They were essentially a crude analog for a human body. Hen eggs did not make the transition to early biological rocket flights due to their fragility, but other types of eggs (sea urchin eggs, fruit fly eggs) were standards of these payloads too." Eggs contain the most basic ingredients for life, so it made sense. If eggs can't survive, what hope have the rest of us?[4]

The other proxy for human life was human tissue. A summary of more than fifty payloads sent up from the late 1940s to late 1950s by Swiss and US scientists includes numerous tissue samples from animals and humans. These payloads included HeLa cells—that is, an immortal cell line cultured from the cervical cancer of Henrietta Lacks, a Black woman who later died of the disease and who was not asked for her consent to their use. The payloads also included another piece of cargo close to my heart: "pieces of skin removed aseptically with a dermatone [*sic*; a dermatome is basically the surgical equivalent of paring knife or potato peeler] during plastic surgery following amputation of the breast." Given the era, I suspect this nameless breast cancer patient was not asked for her consent either.[5]

Eventually, researchers determined that cosmic rays do not cause instant death in living subjects and began to figure out which humans to send into space. In the United States, they chose who you would expect, especially given, according to Bimm's research, that ex-Nazi, former Luftwaffe scientists the government brought to the United States after World War II worked on the space program. These guys had—shocker—specific ideas about what types of bodies should be sent into space, choosing "the healthy, white, male, military test pilot with a degree in engineering," per Bimm.

What I see hidden in the stories of the high-altitude balloons is a tale as old as time. To send men on an adventure, we chopped up the primal mother and sacrificed her—eggs, cervix, and breasts—first. Perhaps, technically speaking, the first human in space was not a white man at all but the cervical cells of Henrietta Lacks; her cell line didn't only go up on the United States' high-altitude balloon experiments but aboard the Soviet Union's *Sputnik-6* as well.

◉

Chicken eggs first went to space in the stomachs of astronauts.

You too can be as constipated as a preflight astronaut if you nosh on a "low-residue" breakfast of steak, eggs, toast, and black coffee. According to the Freedom of Information Act requests filed by the nonprofit site MuckRock, the CIA developed this "top-secret anti-poop diet" for pilots undergoing ten-hour high-altitude spy missions; the first men in space had similar reasons to minimize defecation during flight. Eventually, the meal developed into a launch-day space tradition enjoyed by astronauts, ground crew, and space hipsters.[6]

Eggs are nutritionally dense—they are excellent sources of protein and fat as well as vitamins and minerals—but they are also reminders of home. It's no surprise that they still appear on mission menus. In 2016, European Space Agency astronaut Tim Peake made a video of himself "cooking" breakfast aboard the International Space Station. He produces a packet straight from my personal egg nightmare, a wrinkly, shrink-wrapped yellow blob that he plugs into a panel of buttons, where he injects it with warm water. Then he massages it with his fingers. "In about five minutes, it will be ready to eat," he tells the camera. Where is the romance? The hot pan? The butter? The verbal gymnastics with co-chefs? I'm sure he is happy to be eating anything at all in space where restaurants are scarce. But I wouldn't call it living.[7]

The United States tackled the problem of incubating eggs in space in a uniquely American way, leveraging the power of capitalism. It started in the 1980s with a science teacher who asked a talented eighth-grade student, John Vellinger, to enter a NASA contest to propose an experiment for a space shuttle. Vellinger's family raised chickens in their backyard, and he'd noticed that chickens rotate their eggs, presumably to counteract the effects of gravity, which pulls the yolk downward. He wondered how eggs might develop in a place without gravity and built an incubator prototype out of a wooden box. Although he won the regional science fair, his experiment did not win the national competition. Rather than giving up, he continued to refine the concept, and on his third try, it became a national competition winner. That meant he had a shot at putting his experiment on a real shuttle.

NASA tried to hook winners up with a corporate sponsor to develop their prototypes. For Vellinger, that meant a trip from his home in Indiana to Louisville, Kentucky, to meet with employees of a certain famous fast-food chicken restaurant. The Kentucky Fried Chicken executives agreed to the sponsorship, and suddenly Vellinger's experiment started to become a reality. By this time, Vellinger was a freshman in college studying mechanical engineering. He called it "a dream come true for me." Through Kentucky Fried Chicken's involvement, Vellinger met Mark Deuser, who worked for the chain. As Deuser explained, the fast-food giant has a large research and development department filled with different types of ovens and lab equipment. They use it to develop technology for cooking in their restaurants around the world. Of all the groups at Kentucky Fried Chicken, his was the most logical to help John build the incubator.[8]

They started developing prototypes in the R&D department basement. They had their challenges: blasting an egg into space poses many difficulties. Getting unbroken eggs into space meant protecting them from the force of acceleration during rocket launch, as well as from vibrations that could kill developing chicks. Vellinger called their solution to the problem—a cradle to dampen the effects of vibration and acceleration—their "biggest accomplishment."

Assuming the eggs arrived in space intact, their incubator had to maintain optimal temperature and humidity in the dry atmosphere inside the shuttle. They cracked the humidity problem with baggies and sponges. If the air grew too dry, the crew could periodically moisten a sponge and place it inside the incubator as needed. If the air proved too humid, an air pump could remove moisture. Throughout this process, Vellinger said, they hatched a lot of chicks in the basement of that Kentucky Fried Chicken lab. They also did reliability testing to ensure their gear wouldn't break due to being used in space.[9]

Assuming the gear worked, Vellinger and Deuser had to decide what would go in the incubator's thirty-two slots. They wanted every spot to count, so Vellinger boned up on embryology with a college course and drew on advice from KFC's suppliers. He told me, "I was the one that actually candled the eggs before flight." Egg candling, a technique used since ancient times, employs light—a candle in the early days, and later electric lamps—to reveal what's under the shell. After some discussion they decided to send some embryos that were two days old and some that were nine days old to study the effects of microgravity at different phases of development. Back on Earth, they'd keep an identical set of eggs in an identical incubator. Both the space and control sets would be turned each day.

Their first experiment began and ended on January 28, 1986, when the *Challenger* exploded. Deuser, who came aboard to check on the experiment, was one of the last to leave the craft before lift-off. He and Vellinger watched the tragedy unfold firsthand. (At that very moment, a three-year-old Jordan Bimm, future space historian, was watching it on TV. It is his first memory.)

Three years later, Deuser and Vellinger had a second chance to send their experiment up on STS-29, space shuttle *Discovery*, which launched in 1989. The incubator again held thirty-two eggs, evenly divided between two-day-old and nine-day-old embryos. At the launch site, Vellinger watched the shuttle with his eggs go up. "It was surreal," he said, "just to see your dream that you've worked on for so many years. To see it be actually enabled was super satisfying, super exciting, a real sense of accomplishment that all this time and hard work was paying off. My father was actually there to view the launch—my father gave me a big hug and started crying."

Eventually, the results came in. The two-day-old embryos all died in different stages, proving that gravity did play a role in early development. Most of the nine-day-old embryos survived. Although some of them were dissected while still in the shell, researchers also hatched several, all for the sake of science. The first one to hatch was dubbed Kentucky and sent to live out its days at the Louisville Zoo. The other hatchlings ended up at city zoos in places like Chicago and Lafayette.

Vellinger and Deuser went on to create a business together, Techshot, which develops scientific gear for use in orbit. They have sent more than a few eggs into space at this point, along with other forms of technology, like a bone densitometer, which measures bone density. They are also working on a bioprinter that could use stem cells from humans (or chickens or eggs for that matter) for 3-D printing. The goal is to do 3-D printing of human organs in space for transplant. Imagine needing a kidney and being able to take your own genetic material and make one: that would be revolutionary. It must be done in space, naturally, because like melted plastic set into a mold, the organic material needs some time to toughen up. On Earth, it would simply collapse under its own weight. In 2020 they bioprinted a human knee meniscus in space. The following year, the commercial space infrastructure company Redwire acquired Techshot, where it joined seven other recently acquired space companies.[10]

◉

On a purely selfish level, I hoped the results from the eggs-in-space experiments would answer scientific questions vital to me and to anyone else with ovaries. The organs do more than simply house eggs: ovaries release various forms of estrogen and progesterone, hormones that regulate the reproductive cycle, but also perform many other important functions, like activating osteoblast cells inside bones. Osteoblasts help form new bone cells and control the mineralization of bone. Since estrogen activates them, it makes sense that when estrogen levels drop during menopause, the risk of osteoporosis rises. That goes double for people who hit menopause early. More time spent in menopause means more time for your bones to lose their minerals and become brittle. (I confess, I find it odd to be a woman of a certain age, what doctors would call an "advanced maternal age" if I were pregnant, but whose menopause, caused by an oophorectomy, is considered "premature.") Astronauts are like folks in surgical menopause: they experience abrupt bone loss. Well, in their legs at least. They also tend to gain bone in their forearms, probably from using them so much.[11]

Over the years, Dr. Steven Doty, a senior scientist at the Hospital for Special Surgery, a researcher whose work explores treatments for bone, cartilage, and connective tissue problems,

worked with Deuser and Vellinger at Techshot to send quail eggs
into space to study the problem of bone loss. In addition to people in menopause, people on bedrest also suffer from bone loss
from inactivity. Astronauts spend plenty of their time in orbit
exercising to help counteract the effects of microgravity. It's
not exactly clear what happens to bone density throughout the
astronauts' bodies, as NASA doesn't release much information
about their health. But, as Dr. Doty said, " 'If you don't weight-
bear, you lose bone,' is a standard hypothesis." Doty also said
that the brief info he'd seen on folks who had been in space for
extended times suggested that astronauts returning to Earth
began to regain bone as they started to bear weight again but
that this process stopped after about six months. "So their bone
quality and quantity didn't come back," he said.[12]

Dr. Doty and Techshot saw this issue as an opportunity to use
Techshot's two-tiered incubator-centrifuge, which held thirty-
six Japanese quail eggs. The equipment was compact and self-
regulating—basically set it and forget it—which is key when
astronauts have many duties. Techshot's gear could even inject
fixative into the eggs to preserve them at a particular stage of
growth. The two tiers of the centrifuge allowed for interesting
comparisons. For example, Techshot's and Dr. Doty's collabora-
tion kept two sets of eggs in space, one in microgravity, the other
spun in the centrifuge to simulate Earth's gravity. One of their
questions was whether quail embryos in microgravity continue
to develop a skeletal system and, if so, whether the skeletal system is normal. They hoped it could lead to some insights about
how human bone fractures heal too. Another investigator on
the team, Dr. David Dickman, at Washington University's Central Institute of the Deaf, would investigate the development of
vestibular systems in the ear and their role in vertigo. Unfortunately, a combination of high g-force, heavy ion radiation from

space, and vibration killed about half of the samples, making meaningful statistics impossible. The four embryos that did survive the return trip to Earth showed significantly less bone formation than the ground controls.[13]

For Techshot, it may have all started with an egg, but now their bone densitometer has completed 156 scans while orbiting Earth during several investigations exploring bone loss and muscle-wasting diseases and testing an antiosteoporosis drug in space mice. Still, as of 2021, trends in space research have drifted away from using animals because they require considerable crew time and attention, which are scarce resources. Dr. Doty explained that microgravity experiments in the fields of chemistry, engineering, and physics represent the most "promising" use of spaceflight.

Talking to the researchers filled me with optimism. Vellinger and Deuser spoke casually about life on Mars as if it were a near surety. The use of the word "space" before pretty much any other word still dazzles me. Space eggs. Space embryos. Space research. And yet, despite their optimism, from my layman's armchair here on Earth, the results don't sound too hopeful. Chicken embryos don't develop well in space unless they are already a few days old, which bodes ill for space sex and space birth, as well as for renewable sources of space protein. (On the other hand, bioprinting cloned steak in space, which Techshot was investigating at the time of writing, might work.) Astronauts lose bone mass, which bodes ill for space bones and those in space menopause. Obviously, I'm no rocket scientist, but these obstacles feel insurmountable to me, at least to the limited extent of my knowledge. A space-clone kidney, uterus, or liver, on the other hand, gives me hope for medicine on Earth.[14]

We will certainly continue sending sacrifices into the skies:

offerings to science, to optimism, to some god we pray will save us from climate catastrophe. But in the meantime, perhaps the environmentalists have been right all along: we have only one Earth. All our eggs are on this one planet, so we better stop being bastards and start treating it right.

11

EGG CURES

Humans still raise flocks of sacred chickens, just as the Romans did. Priests no longer tend them, nor do we use them to divine the future. Instead, we rear them in undisclosed locations protected by bodyguards. The number of chickens the US government keeps and what it pays for them are unknown; these numbers are a national security secret but are probably very large. In 2017, the US Department of Health and Human Services paid a single egg provider $42 million for a three-year contract. CNN estimated that the United States might require 140 million eggs each year to fight the flu alone. A nation with a healthy workforce—a vaccinated workforce—thrives, and we owe many of our modern vaccines to the egg.[1]

Humans used to die of contagious diseases in droves; we were desperate to stop the waves of death and disfigurement. Small-pox, for example, once killed about one in three patients. Possibly as early as 1000 CE in China, doctors began blowing powdered smallpox scabs up patients' noses, while later, in Europe, people rubbed smallpox pus into wounds on healthy people's arms to help confer immunity. Only 2 to 3 percent of patients thus treated—variolated—with live disease died from the treatment, a vast improvement over the numbers who died from the disease. Unfortunately, the variolated sometimes kicked off epidemics.[2]

In 1798, British doctor Edward Jenner made a leap forward for smallpox prevention when he figured out that the dairy workers' tales were true: people who caught cowpox, a disease present in livestock, gained immunity to smallpox. No one knew conclusively that smallpox and cowpox viruses were related and that a virus that evolved to attack cows could leave humans unscathed. Jenner had received childhood smallpox variolation, so was unable to experiment on himself. He took material from milkmaid Sarah Nelmes's cowpox sore and injected it into eight-year-old James Phipps. After the boy recovered from feeling a bit unwell, he tried variolating the boy with live smallpox. The vaccine worked. Further tests showed that those inoculated with cowpox, or sometimes horsepox, didn't spread smallpox and were much less likely to die from the treatment. And the vaccine could be transferred from arm to arm, as those inoculated with cowpox developed sores at the injection site. Jenner had made a breakthrough—the creation of a nonlethal, effective live inoculation that used an animal virus similar to a human disease. Jenner's friend, physician Richard Dunning, suggested the new treatment be named vaccine, from *vacca*, the Latin for cow.[3]

Of course, there were still a few kinks to work out, including access to the virus. Unlike bacteria, which lives happily on nutrient broth in petri dishes, many viruses require living tissue to survive over long periods. Doctors in Europe once dried smallpox lymph fluid (pus from sores) on fabric, then reconstituted it in the next town down the road for vaccinations. But the dried lymph didn't survive long sea voyages. Spain turned to orphan boys to transport the jab overseas, infecting them in pairs for backup across the long sea voyage. In return, the boys received education and adoptive families. In 1803, the Royal Philanthropic Vaccine Expedition set sail to Caracas with twenty-two orphans ages three to nine aboard. The first orphans received their injections, and when the sore that formed at the injection site became ripe, physicians onboard transferred cowpox into the arms of the next pair, and so on, until the boat landed in Venezuela and the single ripe pustule remaining went into the arms of the local population. Over the next few years, the mission managed to vaccinate people in Colombia, Ecuador, Peru, Bolivia, and Mexico, then set sail to the Philippines, and from there to China, vaccinating more than 320,000 people in under a decade. Arm-to-arm transmission, which was also used in places like England, had major drawbacks beyond the exploitation of children. It lacked convenience and could spread blood-borne diseases like hepatitis or syphilis along with immunity.[4]

Living animals were a standard method for cultivating viruses for vaccines. By 1836, English physician Edward Ballard had noticed that over time, arm-to-arm transmission methods became less effective, almost as if the virus was petering out. Building on earlier research, he began recommending mixing vaccine strains with fresh cowpox, with the results injected into cows and then back into people, and so on. Eventually, relying on animals to create vaccine supplies helped tamp down some of the problems with arm-to-arm transmission among humans.[5]

Pathologist Dr. Ernest Goodpasture wanted to find a better way. His viral research came in the wake of the 1918 flu pandemic that ravaged the globe. About a third of the global population contracted the virus, and 10 percent of those—50 million people in total—died. While most lethal for young and elderly people, the flu also targeted otherwise healthy folks of any age. In the years that followed, viral research became a hot and well-funded area. Goodpasture believed that fighting lethal viruses depended on finding cheap and easy ways to cultivate pure strains for use in vaccines. In 1925, he became the professor and department head of pathology at Vanderbilt University.

Two years later, he tasked his new assistant, Alice Woodruff, a Mount Holyoke graduate with a PhD in physiology from Yale, to grow fowl pox in something besides a live chicken. He suggested a fertilized egg. At Vanderbilt, she began as an assistant in the physiology department but moved to the Department of Pathology the following year. She'd spend her next three years there working alongside Goodpasture and her husband, Eugene Woodruff, who was a Yale medical school graduate.[6]

Alice's cleverness changed vaccine production for the next century. After candling the eggs, she placed them in egg cups, marking the location of the embryo and air sac in pencil. Using a tiny pair of scissors, she cut a small window in the eggs. Later, others would use dentist's drills for this. She could then pass a needle through the egg membrane and inject the developing chick with fowl pox virus. Afterward, she sealed up this self-contained laboratory with a sliver of glass, held in place by Vaseline, and returned the egg to the incubator.[7]

Although she maintained a sterile environment throughout the process, frustratingly, her embryos kept going moldy before they became infected. She called in her husband, who worked in a lab down the hall, to consult on this issue. Together they realized that the problem lay in the viral culture itself; they discovered that mold had contaminated it. Eugene traced the mold contamination to the skin of the chicks they'd used to harvest the virus, so he shaved and sterilized their heads as well as the knife he used to collect samples. Then he cultured part of a sample in broth to see if any yeast formed. If it didn't, he knew it wasn't contaminated. He also used tiny needles and pipettes and contrived a way to pick out small grains of pure virus. Both methods worked. One day Alice returned to the incubator to find an embryo—still alive!—with a puffy foot. As she told Greer Williams for his 1959 book *Virus Hunters*, "I can't forget the thrill of that moment when Dr. Goodpasture came into my lab, and we stood by the hood where the incubator was installed and I showed him this swollen claw from the inoculated embryo. . . . One might say it represented the Achilles' heel of virus resistance." The following year she and colleague Gerrit Buddingh also grew cowpox and herpesvirus in eggs.[8]

History has not remembered Alice Woodruff. Only credited in about a third of the sources I read, the historical glory goes to Goodpasture. But he did give credit where it was due: Alice received first authorship on their 1931 paper describing the new technique. She felt Goodpasture had been "over-generous" with the credit as "the work started as a result of his suggestion and continued under his expert guidance." As Williams puts it in *Virus Hunters*, after three years on the job and six months after publishing that landmark paper, "she resigned to become a mother." She and Eugene moved to Michigan for his work on tuberculosis. Though she offered vague hopes to Williams of returning to research after her three children finished school, there is no indication that happened. For Alice, as for many professional women of the 1930s, motherhood meant she left research and that her work on vaccines went mostly unremembered. Forget Shakespeare's imaginary sister—I want to know what Alice Woodruff, Ernest Goodpasture's brilliant assistant, might have accomplished if her era's idea of motherhood hadn't confined her.[9]

Goodpasture's work continued. He figured that a single chicken egg could produce about a thousand doses of smallpox vaccine and ran studies proving these vaccines safe and effective. He made discoveries about the etiology of viruses and explored the egg as a medium for growing bacteria, fungi, and protozoa. He also published papers of scientific utility but ones that for laymen like me, at least, also serve up mad scientist vibes. He grew human and chicken skin in eggs to help study skin infections and put fetal membranes (from placentas) in eggs to explore viral infections in the uterus. One of his papers is titled "Infection of Newborn Syrian Hamsters with the Virus of Mare Abortion"—wild stuff. But none of his later discoveries had a wider reach than his research with Alice Woodruff. Their discovery came in 1931; five years later, Max Theiler had used the egg technique to produce a vaccine for yellow fever. After human trials, it became the universal standard. Egg-based vaccines for other diseases, including chickenpox and the flu, followed. Drs. Thomas Francis Jr. and Jonas Salk used the technique to develop a flu vaccine in the 1940s, an effort that the military funded after nearly forty-five thousand soldiers fell ill and died of flu during World War I (by comparison, fifty-three thousand died in combat). The egg-based flu vaccine still accounted for most flu vaccines administered worldwide in 2021. All the manufacturing is based on the procedure Alice Woodruff pioneered.[10]

Eggs can multiply viruses because they can multiply cells. Alive enough to sustain life, eggs are not as cumbersome to care for as, say, a live chicken. Plus each egg is its own little laboratory, conveniently packaged with nature's defenses against contamination. However, the hen eggs' use in vaccines may soon be obsolete.

I learned about this change from my friend Dr. Ane Marie Anderson, a Norwegian medical doctor pursuing a PhD in immunology. I always picture her in a shaggy, full-body sheep suit, the one she wore at the last gala for experience designers I attended, part of a complex joke. She has a bone-dry sense of humor, and I have missed her in the years since my childbirth and the pandemic made travel impossible. When we chatted, she had recently stepped down from a medical position at one of the largest hospitals in Norway. She worked in its vaccine and testing clinic. Her duties included thinking through COVID prevention measures—she was a member of their Quarantine Council—and providing data to the people making the high-pressure decisions during that time. For example, the hospital had about a week and a half to decide which 10 percent of their twenty-five thousand employees would receive the first vaccine shots. When we talked, eighteen months into the pandemic, with Norway on the brink of reopening, she seemed exhausted. Or perhaps it was simply the first day of her new job, returning to work in immunology and transfusion medicine. A colleague of hers said it was a good job for women since it allowed work-life balance for starting a family. Yes, Ane shot back, it's a good job for me since I am finishing my PhD.

Ane explained that although growing vaccines in eggs proved a major development in its day, the technique is still inefficient, particularly since the flu, like COVID, mutates quickly. The flu virus doesn't have an internal control checking to see if it replicates correctly. "So it makes these sloppy mistakes that are evolutionarily brilliant," Ane said. This means that creating a vaccine against the flu is a continually moving target. Winter is flu season, and researchers have plenty of theories why: perhaps it's that the weather drives people indoors where their germs marinate; maybe the short, cold days wear down our immune systems; maybe it's that cold air holds less water, which means virus droplets float around for longer; maybe it's a combination of all these factors. Since flu is a winter thing, the Northern and Southern Hemispheres have opposite flu seasons. The World Health Organization (WHO) watches each hemisphere during its flu season and uses the observations to guess what will circulate in the other hemisphere. The trouble, Ane said, is the chicken eggs. Growing enough vaccine components to make, for example, 160 million doses of the vaccine—the approximate amount required for the US 2019 flu season—requires a six-month lead time. WHO's work monitoring the flu is ongoing. A board of directors reviews data and makes its best guesses about which flu strains will circulate twice a year—in February for the Northern Hemisphere's vaccine and in September for the Southern Hemisphere's vaccine. Governmental medical bodies—such as the Centers for Disease Control and Prevention in the United States and its analogs elsewhere—watch WHO and make localized recommendations to manufacturers who then get to work. In a good year, the flu doesn't mutate much between the time WHO offers its recommendations and when vaccination starts six months later. In a bad year, fast-evolving viral strains mutate quite a bit. This explains the middling efficacy of the flu vaccine—about 30 to 40 percent effective on average. In a bad year, that number might drop below 20 percent, while in a good year it might be higher than 50 percent. Although those numbers might seem low, getting a flu vaccine is still wise, Ane reminds me. The CDC agrees: flu vaccines reduce the chance of hospitalization and death from the flu significantly.[11]

Egg-based vaccination may have a few other problems. "There are mysteries," Ane said. Chickens and humans have some biological differences, to put it mildly. A virus that adapts to thrive in a hen could evolve so that it is unable to bind to human cells or cause human disease. As Ane told me, "That's a really cool and useful general principle," since that adaptation attenuates the virus for humans. But when outside of human cells for too long, certain strains of flu can develop mutations that cause them to adapt to the egg environment so much that they become less effective as vaccines in humans. Case in point: a 2019 paper in the *Journal of Virology* passed a bronchitis virus through eggs a hundred times (and repeated this experiment four times) and compared viral RNA at the end of the process. Its results "highlight the unpredictable nature of attenuation by serial egg passage and the need to develop mechanisms . . . for the next generation of effective vaccines." In the world of vaccinations, the age of egg as bioreactor may be ending, possibly good news for those with egg allergies, although vaccines manufactured via cell culture or recombinant technology have long been available.[12]

A new method for making vaccines is gaining traction, though. In recent years, the cost of synthesizing DNA and RNA has dropped thanks to dramatic technological advancements. To understand how DNA- and RNA-based vaccines work, let's return briefly to the workings of traditional vaccines. Basically, a vaccine alarms the body's immune system in a nondangerous way, which causes the immune system to start preparing its defenses against a particular invader. Traditionally, vaccines injected a body with a live, weakened, or chopped-up viral protein known as an antigen, which triggers an immune response. If the flu is an army of Lex Luthors, then the flu vaccine sends in weakened Lex Luthors—say, his nicer brother Tex Luthor, dead or cryopreserved Lex Luthors, or maybe just Lex's head or a couple of arms. The body uses these weakened versions to set up an antibody training camp that will produce an army of Supermen ready to crush the threat. Gene-based vaccines, on the other hand, use a different form of delivery. Instead of sending in a bunch of Lex's heads, a gene-based vaccine hands out recipes for making Lex's head, in the form of bits of DNA or RNA. And that is what the new tech allows scientists to do: custom-build sequences of DNA and RNA in a lab with a high degree of precision and without as much lead time. In 2021, vaccination campaigns for the worldwide COVID pandemic provided a massive proof of concept for the speed, economy, safety, and effectiveness of this method.[13]

While vaccine production may move away from the egg, as a single-serving, self-enclosed, naturally sterile laboratory, the egg still has uses at the cutting edge of medicine. Consider the chicken as a little machine that turns food into easily accessible proteins and fats, molecules that are useful in the pharmaceutical industry. Designer proteins are the trend for future medicines because they perform a variety of important duties within cells, especially messaging functions. These functions can be useful levers for new medicines to work on, as my computational biologist husband tells me, but the designer proteins that do the work are expensive and hard to create in labs. What if instead we could program the chickens to produce eggs containing drugs?

It sounds like science fiction, but it's already happening. Researchers are using eggs to produce human monoclonal antibodies. A monoclonal antibody is an antibody derived from a single specialized cell. To understand how they work, you have to understand a bit about how the body produces antibodies. Ane calls this process "magical and amazing."

You have up to ten billion B-cells in your body. Each of these white blood cells contains a random receptor on its surface. Picture B-cells as an enormous crowd of master chefs, and antigens, threats from outside the body, as picky customers with highly specific tastes. Each master chef specializes in exactly one egg dish and offers about one hundred thousand servings of it. Inside the lymphatic system, master chefs meet a sampling of customers; the sheer number of chefs practically guarantees that one of them will be holding the correct, precisely calibrated breakfast. When the antigen grabs the bait, the B-cell undergoes an amazing transformation. It rapidly begins to clone itself, creating an army of short-order cooks who rapidly churn out the egg dish (antibody) in question, flooding the body's system with them. (Some of the clones become memory cells, culinary professors who stick around the body for years, in case the antigen pops up again.) Produced en masse, the antibodies bond to specific sites on specific viruses. Viruses are hungry customers: while the average human has but one mouth to fill with egg, viruses are covered with receptors. This means the antibodies can bond to multiple sites at once. I envision the resulting antibody-encrusted viruses like a diner patron coated in egg—both obvious to everyone else at the restaurant and too slippery to do anything useful. Such viruses struggle to bond to host cells, and all the antibodies make them obvious to the sanitation workers of the immune system—macrophages, which engulf and neutralize them. A polyclonal antibody response—the most common type of immune response—is produced by several B-cells at once. Monoclonal antibodies, in contrast, derive from one particular B-cell and are usually made in a lab. Scientists select that particular B-cell for the precise ability of the particular egg dish it cooks.[14]

Monoclonal antibody therapy briefly bolsters a person's immunity to a particular disease and gives a person passive immunity. Whereas vaccines prompt an active immune response, encouraging the body to make its own antibodies, monoclonal antibody therapy injects a person with the needed antibodies directly. It's the equivalent of giving a person some fish instead of teaching them to fish. Since the body hasn't learned to make the antibodies on its own, the effect is temporary. A fetus receives nature's version of antibody therapy. If you jab mom with a flu virus, she will produce antibodies that can cross the placenta to protect her fetus from infection. Likewise, babies receive antibodies from breast milk. Monoclonal antibody therapy is helpful to people who have weak immune systems that can't recognize viruses or a vaccine's antigens. Cancer patients may have immune systems damaged by their treatment; the elderly have aging bodily equipment and likely some comorbidities; and babies and children aren't yet fully developed.[15]

Monoclonal antibodies are useful, but they're also expensive to make because the process involves cloning white blood cells in a lab. This is where eggs come in, leveraging the power of the hen's immune system. A vaccinated chicken passes antibodies into her egg yolk to protect her developing chick. Researchers can then purify the antibodies from the egg yolk and put them to a variety of uses like fighting gum disease, norovirus, the flu, hepatitis, rotavirus, Zika virus, Ebola, and even COVID-19.[16]

But that is only the yolk. The egg white has even more mind-blowing possibilities. Egg whites are old medicine. The ancient Greeks used them to help seal wounds. According to a paper in *PLoS One*, in Asian countries, the membrane under the shell "has been used as an alternative natural bandage on burned and cut skin injuries for more than four hundred years." Recent studies bear out these beneficial qualities. Researchers are studying egg membrane as a salve for burn victims, a material to promote corneal wound healing, a joint health supplement, an antiaging substance, and more. But the most astonishing medical development by far is the use of transgenic chickens to produce drugs. It sounds nuts, yes, but genetically altering a chicken might cause it to create eggs full of pharmaceuticals.[17]

The story begins with researchers Jennifer Doudna and Emmanuelle Charpentier who together won the 2020 Nobel Prize in Chemistry for their 2012 work discovering what is known as CRISPR-Cas9 gene-editing technology. Broadly speaking, viruses have a superpower: they can splice new lines of code into cellular DNA, which leads host cells to generate more viruses. Doudna and Charpentier figured out how to leverage that power. What if, instead of splicing in its own DNA, the virus inserted code chosen by researchers? Doudna and Charpentier's technology opens a vista of possibilities. Cancer arises due to errors in genetic code at the cellular level. Could CRISPR be used to correct that code, leading to remissions? The applications of the field are moving fast. As I wrote this chapter, for example, news broke that scientists had used CRISPR to restore partial sight in a small number of people with Leber congenital amaurosis, whose limited vision arose from DNA errors in the retina.[18]

CRISPR is making sci-fi dreams possible, including that of designer chickens with curative egg whites. In 2015, the Food and Drug Administration approved the first transgenic chicken and eggs for medical use for lysosomal acid lipase deficiency, a rare inherited disorder that causes fat to pile up in various internal organs and the vascular system. Infants with the disease die quickly, while the form that affects adults causes a range of maladies, from liver problems to cardiovascular disease. Scientists used CRISPR to insert a sequence into chicken DNA that codes for a protein, in this case an enzyme that breaks down fat within cells. The chickens then lay eggs with whites full of the enzyme, which is refined and purified, then marketed as the drug Kanuma, and delivered intravenously to patients.[19]

More treatments are on the horizon, waiting for translation to clinical practice. In 2018, a team of Japanese and Korean scientists created transgenic chickens that laid egg whites with anticancer antibodies. The following year, Scotland's Roslin Institute—most famous for cloning Dolly the sheep—produced a transgenic chicken line that laid eggs full of immune-stimulating proteins with antiviral, anticancer, and tissue repair properties. Scientists estimated that the method is economical, with three eggs producing a clinical dose. Meanwhile, Japanese researchers also ran a proof-of-principle study in 2020. They produced eggs containing anti-HER2 monoclonal antibodies, a key ingredient in existing drugs that fight a common but aggressive form of breast cancer. The press has given the whole field, which encompasses other designer animal–based therapeutics, a groaner of a name: farmaceuticals.[20]

The egg's utility has its own dark side, though. Many more female animals will spend their lives never raising a single offspring, the magic of their bodies stolen for another species' gain. On an industrial scale, the production and use of eggs represent the theft of the female body's labor. I find this existentially sad, even while I continue to eat eggs and the idea of next-gen cancer treatments fills me with hope.

12

HUMAN EGGS

The world's most pervasive egg-layers might hold the key to helping screen, treat, and prevent a lethal disease that runs in my family. I find this fact more than narratively convenient.

Human and chicken ovaries are not as different as you might think. Both species have two of them, although in chickens only one is active—typically the left, as in most birds—and a ripe ovum moves into the reproductive tract in the same way it does in humans. A follicle ripens, then bursts through the surface of the ovary like a minuscule wrecking ball, leaving a tiny rip in its wake that the body must repair. In humans, the surface of the now-empty follicle transforms into the corpus luteum, a mass of cells that releases hormones needed for pregnancy.

These similarities make chickens useful to researchers studying ovarian cancer in humans. Cancer can arise in several places within the ovary—in the cells that produce the eggs, the structural tissue that holds the ovaries together, or the epithelial tissue that surrounds the ovaries. The latter is the most common form of ovarian cancer by far, accounting for some 85 to 90 percent of cases. One working hypothesis about why the epithelial tissue is particularly vulnerable has to do with the way ovulation injures the ovary's surface. The body works to repair these injuries, initiating a process of cell proliferation. Sometimes the microscopic construction crew patching up the hole in epithelial tissue acquires an overzealous supervisor who orders them to repair, and repair, and repair—which leads to uncontrolled cellular overgrowth, also known as cancer. Another hypothesis suggests that ovarian cancer starts when a cancerous or precancerous cell from the fallopian tube migrates to the surface of the ovary. Since that surface is the site of so many repair crews, the malignant seed has plenty of opportunity to sabotage the body's natural repair processes. Whatever the mechanism, the evidence shows that the more times a person ovulates, the higher the risk of ovarian cancer.[1]

The roots of cancer lie in cellular DNA. It's a bit more complicated than this, but broadly speaking, we can think of parts of DNA as issuing instructions to the cell to repair itself. This works out well when the DNA gives good instructions, but over time, DNA within particular cells can mutate. These mutations can be caused by environmental exposures or simple errors in replication. The older one is, the more one's cells have copied themselves and the more chances errors have to creep in. Cellular DNA has some really important bits that regulate cell division. For example, some genes spur cellular repair, acting a bit like the gas pedal on a car, while others tell the cell it's time to stop, working a bit like a brake. Mutations can cause the gas pedal to get stuck in the accelerate position, can render the brake useless, or both, leading to unchecked cell replication. Some people—including many in my family—are born with critical pieces of DNA that are already broken. However the errors arise, once they reach a critical mass, the mutated DNA goes rogue, issuing bad instructions that cause unchecked cell growth.[2]

In the case of ovaries, this means the chicken or person can get cancer. The hypothesis that epithelial ovarian cancer arises from rips during ovulation would explain why hormonal birth control, pregnancy, and breast feeding reduce ovarian cancer risk. All of them temporarily pause ovulation, which means fewer rips and therefore fewer opportunities to repair the surface of the ovary.[3]

If that theory is true, then it makes sense for hens to be at high risk of developing ovarian cancer. Chickens ovulate about once a day during peak laying periods. Furthermore, they pump all their eggs out of a single ovary. Compare that to humans, who ovulate once a month and may do so from either ovary. By the time a chicken is two or three years old, it has ovulated as much as a menopausal woman but over a much shorter period and all from a single gonad. The frequency of chicken ovulation adds up: while people with ovaries have, on average, a 1.3 percent chance of developing ovarian cancer across a lifetime, in chickens that number climbs exponentially. On average, depending on which statistics one consults, chickens have a lifetime ovarian cancer risk of approximately 10 to 35 percent, or by some calculations as high as 83 percent.[4]

Chickens provide an excellent experimental model for the disease in humans because of these similar traits. In addition, ovarian cancer in chickens progresses similarly to the disease in humans. And to top it off, humans can control every aspect of chickens' lives to make sure experimental results are based on compliance. Scientists have done studies in chickens to explore various methods for preventing ovarian cancer, including injecting them with birth control, which reduced ovarian cancer by 15 percent; putting them on starvation diets, which reduced the disease by a factor of five; feeding them a diet that was 10 percent flaxseed, which significantly reduced the severity and incidence of ovarian cancer in older (3.5-year-old) hens; and giving them vitamin D treatments, which did nothing. We already knew birth control works to lower the risk of ovarian cancer in women, but—modern diet culture notwithstanding—starvation diets are unlikely to become a popular method of cancer prevention. The flaxseed poses an interesting option, though. For humans to eat a proportionally similar amount of flaxseed seems untenable, although a randomized controlled trial for a different disease showed that in women with breast cancer, eating 25 grams of flaxseed a day in a muffin helped control tumor growth, which is promising. Based on the hen and breast cancer studies, in 2015, Southern Illinois University began a small clinical trial in ovarian cancer patients in remission, in which they ate 20 grams of flaxseed—a couple of tablespoons—every day. Unfortunately, the study did not yield results, as they were not able to enroll enough patients.[5]

Other studies have explored biomarkers in chickens with ovarian cancer that are also present in many types of human cancer, for example, the protein CA 125, for which my doctor regularly screened my blood. Researchers are also using chickens to explore other proteins known to regulate cancer growth, such as EGFR, HER2, p53, and TGF-alpha. The latter is a protein my husband, a bioinformatics scientist, studies in human cancer. Chicken research has also yielded discoveries about proteins such as SERPINB3, which may play a role in cancer development. Translated to human patients for a study in *PLOS One*, the research provided evidence that in women with ovarian cancer, the presence of SERPINB3 may predict resistance to certain kinds of chemotherapy. This is all good news for cancer patients, as it's generally agreed that chemo sucks and taking chemo that doesn't kill your cancer sucks doubly.[6]

The hen's high likelihood of developing ovarian cancer and her utility in research makes me feel tenderly toward her. On my mother's side, many of us share an inherited mutation on one of our *BRCA1* genes. Every person has two copies of this gene—one from each parent—in every cell of their body. *BRCA1* regulates the repair of double-stranded DNA breaks, which is an important cellular job. If something takes out a cell's working copies of *BRCA1*, the normally orderly process of cell division can begin to go awry. Over time, errors in DNA accumulate, which can lead to cancer. Some people, including me, inherited a faulty copy of *BRCA1* in every cell of our bodies, which dramatically raises our risk of breast and ovarian cancer. The numbers are not in my family's favor: a 55 to 72 percent lifetime chance of developing breast cancer and an ovarian cancer risk more on par with the average hen—a 39 to 44 percent lifetime chance, about a thirty-fold increase compared to the average person with ovaries. Adding to the fear, cancer in *BRCA1* patients tends to be more aggressive and strike at a younger age than cancer in the general population.[7]

The family pathos around cancer is wide and deep, big enough for me to have written the book *Pandora's DNA* about it. I don't want to recap the whole story here, so let's just say I've got a lot of family history. If you look at my grandma's generation, three women experienced six total bouts of cancer (four breast, two ovarian)—unevenly distributed among my grandma and her sisters—and two cancer deaths. My mom developed breast cancer at thirty, when I was a toddler. Even before researchers put a genetic face on our family suffering with the discovery of *BRCA1* in the early 1990s, relatives began removing healthy body parts because the family curse was so clear. After the identification of *BRCA1*, removing body parts became part of the medical recommendations for people like us. You can't get cancer in organs you no longer have. So I opted to remove my healthy breasts when I was twenty-eight and then my ovaries at thirty-nine, while writing this book.[8]

Through the absence of my own eggs, I've come to understand them better and perhaps to love them even more. Generally speaking, removing a person's ovaries isn't advisable. Undergoing the "Change" before age forty-five comes with a host of risks, including mood disorders (depression and anxiety), body image issues, sexual dysfunction, compromised self-confidence, and increased risk of conditions like cognitive impairment and dementia, osteoporosis, heart disease, and metabolic syndrome. In fact, ovarian removal raises the likelihood of all-cause mortality—that is, death for any reason. For the average person, as one study bluntly put it, "At no age was oophorectomy associated with increased overall survival." And yet. And yet. I am not the average woman. I am a mutant with a high risk of ovarian cancer. For *BRCA1* carriers like me, doctors recommend the procedure between the ages of thirty-five and forty, or "after childbearing is completed," as the surgery prevents us from dying young of ovarian cancer. I found the procedure more attractive than facing the disease that spread to my great-aunt's spine, rendering her paraplegic, killing her slowly as her children watched.[9]

Removing the human egg-creating apparatus has serious consequences. Ovaries do far more than simply churn out the potential for new life. They also regulate metabolism in a thousand ways that science hasn't yet mapped to my personal satisfaction. We do know that ovaries release three types of estrogens as well as progesterone, hormones that control both fertility and metabolism, as well as performing functions in organs like the liver, heart, bone, brain, and others. During an average perimenopause, a person's hormones ramp down over a period of years—time estimates vary from four to fourteen—but continue to release diminishing levels of hormones in perpetuity, a little cushion against the slings and arrows of menopausal fortune.

If you remove the ovaries surgically, the whole process takes hours—hours!—instead of a decade, and if you don't start on hormone replacement, levels of estrogen and progesterone rapidly plummet. Going into the operation, my biggest fear was that I'd go to bed Snow White (OK, a middle-aged, tubby Snow White) and wake up the witch.[10]

Reader, it happened.

My hourglass shape has turned into more of a clock, as predictably fat shifted upward. Regular issues aside, my abrupt menopause had a few surprises in store for me. Estrogen apparently helps regulate how one's calcium carbonate crystals attach to one's inner ear, which has implications for balance. I spent a few months with extreme vertigo, until repeated maneuvers at the physical therapist set me right. It's also like having the worst parts of puberty all over again. For a while, I had the appetite and acne of a teenage boy but the metabolism of a sixty-year-old woman. Apparently, estrogen interacts with the hypothalamus, which helps regulate satiety and fat distribution. This may help explain why surgical menopause comes with increased risk of cardiovascular disease and metabolic syndrome. It may also explain why, postsurgery, I have experienced intense sugar cravings, as well as my first blood test to ever show high cholesterol.[11]

Hormone replacement therapy, which I began the day of my operation, helped, but it wasn't simple to get right. I had my first hot flash on Christmas morning while I was trying to orchestrate the festivities. Stress, apparently, decreases estrogen in the body, which can heighten menopausal symptoms. I began to feel like a drug addict, always jonesing for more estrogen. When my initial weekly estrogen patch left me with a mild rash under the adhesive, I asked to try something else. My menopause specialist suggested the rash wasn't so bad but, surrendering to my wishes, prescribed me a smaller, twice-a-week version. That's how I learned that I'm sensitive to hormones and their formulations. Every three days I shuttled between high, medium, and low estrogen states, that is, nausea, happiness, and weeping-on-the-floor depression. My hair began falling out by the handful, so much so that one day, when my spouse walked into the bathroom and saw the strands in my hands, he said, "Oh, that's not good—reminds me of when I was going bald," and gave me a big hug.[12]

The mental symptoms of egg removal are also considerable. I had the problems that are apparently usual during menopause—it's hard to think when your brain is rewiring itself—but my most prominent symptom by far involved my mood. I used to pride myself on a certain stoicism, but now rage overpowered me. It was as if every small insult I had ever swallowed, every questioning of my competence that I ignored, every infuriating setback I powered through came home to roost. I thought I was above anger, but it turned out rage buried itself in my innards, a little gold hoard guarded by the dragon estrogen. It poured out of me at the slightest provocation, a molten river ceaselessly seeking an outlet. In 2021 the *New York Times* ran an article titled "When Your Home Is a Hormonal Hellscape" about the trials of parenting teens while perimenopausal. I laughed bitterly and thought, "Try surgical menopause with a three-nager." I attempted not to yell at my kid, with mixed success. I was new to mastering full-frontal rage and channeling it appropriately. I hired a therapist.[13]

Although I eventually got my hormones stabilized, I still could not expect to live like a premenopausal woman, my specialist told me: change had come. Perhaps surgical menopause's most irritating aspect is that, medication aside, it's best managed by doubling down on all the experts' hard-to-follow advice about clean living. No more late-night martini parties. I too need to lift weights, but becoming a hot body is a pipe dream the literature suggests I ought to leave behind. No, for me, weightlifting is now about staving off osteoporosis and weight gain. All the work, merely to stave off decay. Only a true boss crone is mature enough to eat her vegetables and go to bed early and sober.[14]

I have decided to lean into this whole insta-crone thing. I am living the crone lifestyle, with high-fiber breakfast cereal, large eccentric muumuus, loud jewelry, and being a total badass who is strong AF. Or at least that is what I keep telling myself.

When I think about my eggs, I cannot help but see myself in the hen and the way our culture treats her—as an egg-producing machine rather than a living, breathing individual. I felt like an object during the last part of my pregnancy, when doctors began ignoring my discomfort. "Oh, good," my OB said when I told her about the stabbing pelvic pain that arrived unpredictably. "That means your pubic bones are opening for the birth." Is there anything more feminine than quietly enduring misery? I speak of the ankle-breaking, Achilles tendon–shortening heels; the tight shapewear; the menstrual blood that must be kept hidden; the painful hair removal; and the moment when the nurse tells you not to scream despite the baby's head literally ripping your body open.

◉

When a hen's main ovary becomes disabled or stops producing estrogen, occasionally she changes gender. The medical literature has dubbed this transformation "spontaneous sex reversal." She grows chin wattles and a cockscomb, stands more erect, and starts crowing. Her right ovary sometimes develops into an ova-testes that produces sperm. In other words, while the hen's cellular DNA remains female, her sex expression changes to that of a cock.*[15]

I am reminded of journalist, author, and fellow BRCA carrier Masha Gessen, who also had their breasts and ovaries removed. We have different mutations: mine is in my *BRCA1* gene; theirs is in their *BRCA2* gene, with similar but not identical effects. They wrote *Blood Matters* about it, a book I read avidly when I was making my own decisions. As they told an interviewer in 2019,

> I've actually made decisions about hormone replacement as a blank slate. I went on female hormones at first—I went to estrogen—I mean both female and male. Obviously, both men and women have both estrogen and testosterone, but in different proportions. And I felt horrible. I felt really, really, awful taking estrogen. I thought, "Hell, let me go in the other direction." I started taking testosterone, and I'm having a lot more fun on testosterone than I did on estrogen. I'm on a very low dose, but that has created a nonbinary gender situation for me. But for me, the physical stuff was primary, and then the gender stuff was secondary.

Gessen's surgery to remove ovaries and uterus sparked hormonal changes that brought them to a new realization about their gen-

---

* Chicken gender is fascinating. Cocks with a damaged main testes may begin to exhibit female traits and even lay eggs. Whether these eggs, like the sperm of the hen-turned-cock, are viable is less clear. The species also includes gyandromorphs, birds with gender literally split down the middle. One side of the bird appears like a rooster, the other like a hen. Gyandromorphism can be split left-right or front-back or take the form of a mosaic, with female and male cells studded throughout the body.

der identity. I think about that sometimes, because this shift in hormones and body parts has affected my own sense of gender.[16]

As a young woman, I found my culture's tendency to underestimate me incredibly irritating. I didn't want to be a "young woman." I wanted to be treated as a competent person. I sometimes used to say I was person first, vagina-haver second. Removing my breasts in my late twenties created anxiety around my own womanhood for me. On one hand, breasts don't make a woman, but they are organs that US culture loudly associates with femininity—a physical marker that my oncologist and plastic surgeon replaced with a silicone simulacrum. What did that mean about who I was now? To soothe my anxiety, I adopted elaborate beauty routines and attended risqué parties by and for Nordic experience designers, which left me with wonderful memories. I hadn't planned to live past thirty, so I didn't know what to do with myself in this strange, healthy afterlife. But filling the void with body maintenance and excessive art projects didn't work. I felt purposeless and slipped into a deep depression that continued through the birth of my son, who I now know will be my only child.

With my ovaries gone, I feel somehow that I have returned to my old self: person first, woman second. Maybe my gender has been "crone" all along.

As for spent hens, they are sent to the slaughterhouse to become chicken soup or pet food—or they are simply euthanized and buried in landfills. I find echoes of that treatment in the way American culture deals with menopause, as "a predeath, a metamorphosis from a woman to a crone with her exit ticket already punched," as Dr. Jen Gunter, gynecologist and author, put it in a *New York Times* article. And yet there is power in the role too. Any type of menopause has its confusing moments—where is the middle-school health teacher to show me palm-sweating videos about it in the cafeteria?—but it also has its benefits. In menopause, there is freedom from menstrual blood, from worries about pregnancy, and from pregnancy and childbirth themselves.[17]

Chickens, like most animals on Earth, do not go through menopause. They do, however, experience "henopause," usually around five years of age. During henopause, a chicken either goes long stretches between eggs or stops laying entirely. Animals that receive good care can live ten years or longer. But although henopause turns them into "spent" hens, or "biddies," it's not a metabolic process the way menopause is in humans. The only mammals who share that distinction are three species of whales and orcas—known as killer whales but in the dolphin family—who hit menopause around forty but can live to ninety. And it's the menopausal orcas who end up at the top of a pod's complex social hierarchy. Biologists have a theory about it called the grandmother hypothesis. According to the authors of a paper in the *Proceedings of the National Academy of Sciences*, "By stopping reproduction, grandmothers avoid reproductive conflict with their daughters, and offer increased benefits to their grandoffspring." The most vital of those benefits occurs when resources are stretched thin. In humans, evolutionary biologists theorize that menopause helps children, since their moms are less likely to die in childbirth, and that grandmas often provide care and support to their daughter's children. Menopause is not pre-death but a fact of life. As Dr. Gunter told the Canadian Broadcasting Corporation in 2021, "Menopause is a sign of strength. The fact that we evolved to live beyond our ovarian function and other animals didn't says something about our strength."[18]

I'm all out of eggs. That means I no longer carry the responsibility of not-quite-yet-to-be life. I have been freed from the need for birth control, blood on my jeans, and the terror that laws passed by men in suits might require me to continue a pregnancy I don't want. I am no longer a servant to my body's potentiality, and this too is a change. The irony, of course, is that as this transition has helped me gain clarity about who I am, my culture rarely notices women in this life stage. My mother loves the invisibility of her station because she can go to the grocery store in her gardening gear and no makeup and be as private as she likes. I have not yet reached that level of enlightenment. I want to be seen and listened to and, occasionally, to have my pain witnessed.

Although I miss my eggs, particularly their metabolic effects, I find myself anticipating the next life change with eagerness. I look forward to acquiring more wisdom and to continuing the shift in perspective that parenthood began for me, a decentering of the self, a wide and more gracious capacity to feel joy on behalf of my child but also on behalf of all sorts of people in my life. Already the absence of my eggs has given me narrative distance from my martini-swilling, *Twin Peaks* elevator party years. I can enjoy them as wonderful memories rather than sweat over arranging and topping them with new adventures. And that is also a kind of freedom. I am happy to be alive. I survived death's first assault with a mastectomy that helped me live beyond the family expiration date of thirty. A decade later, after my ovarian surgery, my ninety-eight-year-old grandpa wrote to congratulate me: "Word on the street is that you won a knockout victory in your bout with ovarian cancer." Thanks to my canny mother and grandmother, who did their research, thanks to the grandmother effect, ovarian cancer never saw it coming. I have survived the second wave of death by egg that haunts my family. In a way, I feel born again, just as I did after my mastectomy. This second afterlife demands no hard biological labor from me, no more tangos with the surgeon's knife, no more childbirth, only to love my husband and child and do the artistic work I want to do. I have danced safely through the field scattered with eggs, right through my own cosmic loophole. On the other side, comfortable clothing awaits. So does a collection of omelet pans. And I am already working on a participatory piece that features dancing, a mallet, and a pile of raw eggs.

ACKNOWLEDGMENTS

No book enters the world without a flock supporting the author. I am deeply grateful to my hardworking agent, Jane Dystel, who gifted me with the germ of this idea. Thank you to my editor, Amy Cherry, who stuck with me through a concussion, surgical menopause, and a global pandemic, and whose thoughtful edits greatly improved the manuscript.

Thank you to the team at W. W. Norton—crack editorial assistant Huneeya Siddiqui, amazingly thorough copy editor Pat Wieland, and the marketing and publicity departments for all their hard work.

My fact checkers, Christopher Mele and Kelsey Kudak, helped the manuscript by holding me accountable for its contents. I am grateful to them, and any errors that remain are my own. Thank you to Pat Pholl for permissions help.

Many thanks as well to everyone who sat for an interview. Whether or not you appear in the manuscript, you provided valuable insight and context, without which this book would be impossible: Ken Albala, Ane Marie Anderson, John Bates, Rich Boling, Stephen Doty, Mark Deuser, Mattie Faint, Paul Freedman, Jennifer Gleason, Chris Leahy, J. Kenji López-Alt, Jacques Pépin, Donna Pierce, Amanda Rawls, Alan Rocke, Debbie Smith, Mark Thomas, Richard Thomas, Jeremiah Trimble,

John Vellinger, and Jim Wilson. Thank you to my amazingly helpful correspondents, who shared information about space, witches, Martha Washington, and titanic eggs with me—respectively, Jordan Bimm, Kristen Sollée, Mary V. Thompson, and Carolin C. Young. Thank you to the authors of the many sources cited in this book: I am grateful to stand on the shoulders of giants.

I have a soft spot in my heart for librarians of all stripes. Thank you to Christopher Ryland, curator at the History of Medicine Collections and Libraries at Vanderbilt, for sending me valuable documents and information about Alice Woodruff. Thank you to Eva Patterson, Teressa Snyder, and Dan McMahon of the Marin County Free Library for heroically scanning information on the Farallon eggers for a random author midpandemic.

This book would have greatly suffered without thoughtful reads, encouragement, and feedback from my writing group. Thank you to the Freedmanites—Alice Sparberg Alexiou, Johnathan Englert, Dina Hampton, Elizabeth Kadetsky, Chris Lombardi, Karen Pinchin, Joan Quigley, and Grace Williams—who have enriched my work and my life immensely. Thanks to my coach Richard Nash for the insights that kept me moving in the right direction, even when the world wasn't.

Thank you to friends and associates who gave me story tips, shared favorite recipes, or answered my many random queries about eggs and associated phenomena: Daud Alzayer, Aatish Bhatia, Whitney Beltrán, Sarah Bowman, Fox Harrell, Dominika Kovacova, Shuo Meng, Markus Montola, Jason Morningstar, Alex Roberts, Jeeyon Shim, John Stavropoulos, and many more friends on social media—too many to name, but you know who you are!—who gifted me with links across the writing process or participated in my breakfast egg exchange.

Thank you to my writing buddies, omelet buddies, friends,

and remote "coworkers" for providing moral support, motivation, sounding-board services, and occasionally beta reading: Kellian Pletcher Adams, Lisa Biagiotti, Chip Cheek, Julia Henderson, Katherine Hunt, Kathryn Hymes, Jon Kruger, J. Li, John McManus, Sarah Miles, Danny Mitarotondo, Jason Morningstar, Caro Murphy, Todd Perry, Alex Roberts, Janell Sims, James Stuart, Avital Ungar, Urban Waite, Shannon Ward, Jerry Williams, and Sara Williamson.

Thank you particularly to everyone who kept the home fires burning while I wrote. Thank you, Marconi Sousa, for all your work. Thank you to my amazing parents-in-law, David Locke and Cathleen Read, who cared for my child full-time throughout the pandemic and who have supported me at every step along the way. Thank you to my parents, Dick and Gretchen Stark, for providing backup child care at crucial intervals, for instilling a love of knowledge and the kitchen in me, and for all the eggsperiments. Thank you to George Locke, my loving, supportive life partner, for tolerating so much egg talk and consuming so many experimental results. And thank you to Rog for being a consistent source of joy and delight.

One morning, when I had begun work on this manuscript, I found a live sparrow sitting atop my computer monitor. It had wormed its way into the house through a gap in my office's window air conditioner and had dragged straw up into the ceiling light fixture to make a nest, I think. Its droppings graced my computer and the pages next to it—a little "blessing" from my subject matter. After a brief—and terrifying—interval when it swooped down over my head, my husband used a blanket to shepherd it out an open window. So let me end with a big thank you to nature's female birds, whose labors we do not celebrate enough.

NOTES

PROLOGUE

1. Pia Lim-Castillo, "Eggs in Philippine Church Architecture and Its Cuisine," in *Eggs in Cookery: Proceedings of the Oxford Symposium of Food and Cookery 2006* (Devon, UK: Prospect, 2007), 114–124.
2. Merry Sleigh, "Ovum," in *Encyclopedia of Child Behavior and Development*, ed. S. Goldstein and J. A. Naglieri (Boston: Springer, 2011), 1050–1051, doi: 10.1007/978-0-387-79061-9_2063.
3. Paul R. Ehrlich, David S. Dobkin, and Darryl Wheye, "Eggs and Their Evolution," Stanford University, 1988, accessed March 17, 2022, https://web.stanford.edu/group/stanfordbirds/text/essays/Eggs.html.
4. Alie Ward, "Oology with Dr. John Bates," *Ologies*, podcast audio, August 13, 2018, https://www.alieward.com/ologies/oology.
5. Tim Birkhead, *The Most Perfect Thing: Inside (and Outside) a Bird's Egg* (New York: Bloomsbury, 2016), 171–177; Florence Baron et al., "Egg-White Proteins Have a Minor Impact on the Bactericidal Action of Egg White toward *Salmonella* Enteridis at 45C," *Frontiers in Microbiology* 11 (October 2020), doi: 10.3389/fmicb.2020.584986; Yoshinobu Ichikawa et al., "Sperm-Egg Interaction during Fertilization in Birds," *Journal of Poultry Science* 53, no. 3 (July 2016), doi: 10.2141/jpsa.0150183; Tomohiro Sansanami et al., "Sperm Storage in the Female Reproductive Tract in Birds," *Journal of Reproduction and Development* 59, no. 4 (August/September 2013), doi:10.1262/jrd.2013-038.
6. Birkhead, *Most Perfect Thing*, 121–146.
7. Birkhead, *Most Perfect Thing*, 90.

CHAPTER 1: COSMIC EGG

1. *Kalevala*, vol. 1, trans. John Martin Crawford (Cincinnati: Robert Clark, 1888), 9, publicdomainreview.org/collection/Kalevala.

2. *Kalevala,* 5–18; Barbara C. Sproul, *Primal Myths: Creation Myths around the World* (New York: HarperOne, 1979), 176–178.

3. David Leeming, with Margaret Leeming, *A Dictionary of Creation Myths* (New York: Oxford University Press, 1994), 73.

4. Sproul, *Primal Myths,* 349–350.

5. Malayna Evans Williams, *Signs of Creation: Sex, Gender, Categories, Religion and the Body in Ancient Egypt,* PhD diss., University of Chicago, 2011, 128–153.

6. Andrew Lawler, *Why Did the Chicken Cross the World: The Epic Saga of the Bird That Powers Civilization* (New York: Atria Paperback, 2016), 188.

7. "Punic Wars," *Encyclopedia Britannica,* accessed November 20, 2021, Britannica.com/event/punic-wars; *Scholia Bobiensia,* trans. T. Stangl (1912), accessed November 22, 2021, attalus.org/translate/bobiensia.html.

8. Françoise Dunand and Christiane Zivie-Coche, *Gods and Men in Egypt: 3000 BCE to 395 CE,* trans. David Lorton (Ithaca, NY: Cornell University Press, 2004), 9; Williams, "Signs of Creation," 209.

9. John Hale, *A Modest Enquiry, into the Nature of Witchcraft* (Boston: B. Green, 1702; Ann Arbor, MI: Evans Early American Imprint Collection, 2021), https://quod.lib.umich.edu/e/evans/N00872.0001.001/1:5.14?rgn=div2;view=fulltext, 132.

10. Kristen Sollée, email correspondence, September 26, 2020; Paloma Cervantes, "What Is a Limpia (Spiritual Cleansing)?" Institute of Shaminism and Curanderismo, accessed November 22, 2021, https://www.instituteofshamanismandcuranderismo.com/What-Is-A-Limpia-Spiritual-Cleansing/; Alex Swerdloff, "Cleanse Your Aura with the Power of Eggs," *Vice,* last modified October 28, 2016, https://www.vice.com/en/article/wnbxnn/cleanse-your-aura-with-the-power-of-eggs.

11. Lisa Stardust, "The Future Is Now: Manifesting with the Gemini New Moon," Hoodwitch, accessed November 22, 2021, https://www.thehoodwitch.com/blog/2019/6/2/the-future-is-now-manifesting-with-the-gemini-new-moon.

12. Alan Rocke, phone interview, October 12, 2020.

13. Jordan Bimm, email correspondence, January 2020.

14. Hannah Ritchie and Max Roser, "Biodiversity and Wildlife," Our World in Data, accessed November 22, 2021, https://ourworldindata.org/biodiversity-and-wildlife#how-many-species-are-there; Gifford Miller, John Magee, Mike Smith, et al. "Human Predation Contributed to the Extinction of the Australian Megafaunal Bird *Genyornis newtoni* ~47ka," *Nature Communications* 7 (January 29, 2016), https://www.nature.com/articles/ncomms10496. The fact that eggs are generally safe to eat is only true of animal eggs. Of course, many of our vegetables and spices consist of plant eggs—seeds—but there are far more poisonous

seeds in the plant kingdom than in the animal kingdom. Don't eat random seeds, kids!

15. William J. Stadelman, "II.G.7/Chicken Eggs," in *The Cambridge World History of Food,* ed. Kenneth F. Kiple and Kriemhild Coneé Ornelas (Cambridge: Cambridge University Press, 2000), 500.

16. Makiko Itoh, "The Raw Appeal of Eggs," *Japan Times,* September 16, 2014, https://www.japantimes.co.jp/life/2014/09/16/food/raw-appeal -eggs/; Naomichi Ishige, "V.B.4/Japan," in *Cambridge World History of Food,* 1176; Ishige, "Eggs and the Japanese," in *Eggs in Cookery: Proceedings of the Oxford Symposium on Food and Cookery 2006,* ed. Richard Hosking (Devon, UK: Prospect, 2007), 100–106.

17. Susan Weingarten, "Eggs in the Talmud," in *Eggs in Cookery,* 273.

18. Sephardi Kitchen, "Huevos Haminados (Sephardic Jewish-Style Eggs)," Food.com, accessed November 24, 2021, https://www.food.com/recipe/ huevos-haminados-sephardic-jewish-style-eggs-317802.

19. Kenneth Albala, "Ovophilia in Renaissance Cuisine," in *Eggs in Cookery,* 13; Kenneth Albala, "V.C.2/Southern Europe," in *Cambridge World History of Food,* 1204–1205; Natasha Frost, "How Medieval Chefs Tackled Meat-Free Days," Atlas Obscura, July 27, 2017, https://www.atlasobscura .com/articles/mock-medieval-foods#:~:text=Christians%20 observed%20at%20least%20three,to%20commemorate%20the%20 Virgin%20Mary.

CHAPTER 2: EGG HUNT

1. Barbara Mearns and Richard Mearns, *The Bird Collectors* (San Diego: Academic, 1998), 205–207; Mark Barrow, *A Passion for Birds: American Ornithology after Audubon* (Princeton, NJ: Princeton University Press, 1998), 41.

2. Michael Anft, "This Is Your Brain on Art," *Johns Hopkins Magazine,* March 6, 2010, https://magazine.jhu.edu/2010/03/06/this-is-your-brain -on-art/; Oshin Vartanian, Anjan Chatterjee, Lars Brorson Fich, et al., "Impact of Contour on Aesthetic Judgments and Approach-Avoidance Decisions in Architecture," *Proceedings of the National Academy of Sciences* 110 (Suppl. 2) (June 18, 2013), https://doi.org/10.1073/pnas .1301227110.

3. Mearns and Mearns, *Bird Collectors,* 40–42.

4. Charles Bendire, "Circular No. 30, Appendix: A List of Birds the Eggs of Which Are Wanted to Complete the Series in the National Museum, with Instructions for Collecting Eggs," *Proceedings of the United States National Museum* 7 (1884): 613–616.

5. Sara Wheeler, "The Nice Man Cometh," *Guardian,* November 4, 2001, https://www.theguardian.com/books/2001/nov/04/biography.features1.

6. David Crane, *Scott of the Antarctic: A Life of Courage and Tragedy* (New York: Knopf, 2006), 371; Joy McCann, "Penguins Were a Lonely Explorer's Best Friends," *Atlantic,* April 3, 2019, https://www.theatlantic.com/science/archive/2019/04/penguins-southern-ocean-explorers-best-friend/586189/.

7. Crane, *Scott of the Antarctic,* 447.

8. Apsley Cherry-Garrard, *The Worst Journey in the World: Antarctic 1910–1913* (New York: George H. Doran, 1922), 233–237, 242, accessed November 23, 2021, https://archive.org/details/worstjourneyinwo01cher/page/n9/mode/2up; Sir Ranulph Fiennes, *Race to the Pole: Tragedy, Heroism, and Scott's Antarctic Quest* (New York: Hyperion, 2004), 233.

9. Fiennes, *Race to the Pole,* 236.

10. Robert Falcon Scott, *Scott's Last Expedition (Classics of World Literature)* (London: Wordsworth Editions, 2012), 257, Kindle.

11. Cherry-Garrard, *Worst Journey in the World,* 299; Robin McKie, "How a Heroic Hunt for Penguin Eggs Became 'The Worst Journey in the World,'" *Guardian,* January 14, 2012, https://www.theguardian.com/uk/2012/jan/14/penguin-eggs-worst-journey-world.

12. Tim Birkhead, *The Most Perfect Thing: Inside (and Outside) a Bird's Egg* (New York: Bloomsbury, 2016), 13–15.

13. Joseph J. Hickey and Daniel W. Anderson, "Chlorinated Hydrocarbons and Eggshell Changes in Raptorial and Fish-Eating Birds," *Science* 162, no. 3850 (1968): 271–273, http://www.jstor.org/stable/1725067; John Bates, "Eggshells, DDT, Collections, and Study Design," Field Museum, Chicago, last modified May 30, 2018, https://www.fieldmuseum.org/blog/eggshells-ddt-collections-and-study-design.

14. "Bird Egg and Nest Collections," Natural History Museum, London, https://www.nhm.ac.uk/our-science/collections/zoology-collections/bird-egg-and-nest-collections.html; J. P. Pickard, "The Egg Man Cometh," *No. 5 Regional Crime Squad, Hatfield,* 54 Police J. 279 (1981), accessed May 27, 2020, https://heinonline.org/HOL/LandingPage?handle=hein.journals/policejl54&div=32&id=&page=; Kirk Wallace Johnson, *The Feather Thief* (New York: Penguin, 2018), 111.

15. Mark Thomas, phone interview, January 13, 2020.

16. Thomas, interview; *Poached,* directed by Timothy Wheeler (New York: Ignite Channel, 2015), documentary; "(pounds) 2500 Fine and Jail Warning for Egg Thief," *Herald Scotland,* January 15, 2003, accessed April 20, 2022, https://www.heraldscotland.com/news/11904004.pounds-2500-fine-and-jail-warning-for-egg-thief/.

17. Julian Rubinstein, "Operation Easter," *New Yorker,* July 15, 2013, https://www.newyorker.com/magazine/2013/07/22/operation-easter.

18. Patrick Barkham, "The Egg Snatchers," *Guardian,* December 10, 2006, https://www.theguardian.com/environment/2006/dec/11/g2.ruralaffairs.

19. Thomas, interview; "Slavonian Grebe Facts/*Podiceps auritus,*" RSPB, accessed March 28, 2022, https://www.rspb.org.uk/birds-and-wildlife/wildlife-guides/bird-a-z/slavonian-grebe/; "Red Backed Shrike Bird Facts/*Lanius collurio,*" RSPB, accessed March 28, 2022, https://www.rspb.org.uk/birds-and-wildlife/wildlife-guides/bird-a-z/red-backed-shrike/.

20. "Thank god you've come...," as quoted in Rubinstein, "Operation Easter"; "Norfolk Man Who Illegally Hoarded 5,000 Rare Eggs Jailed," BBC, November 27, 2018, https://www.bbc.com/news/uk-england-norfolk-46358627; Peter Walsh, "Man Jailed and Told to Give His Collection of 5,000 Rare Bird Eggs to Natural History Museum," *Eastern Daily Press,* November 27, 2018, https://www.edp24.co.uk/news/crime/norfolk-collector-daniel-lingham-must-give-his-5-000-rare-1310900; Sam Russell, "Man Who Illegally Collected More Than 5,000 Rare Bird Eggs Jailed for Threatening Population Species," *Independent,* November 28, 2018, https://www.independent.co.uk/news/uk/crime/bird-egg-thief-jailed-daniel-lingham-norfolk-norwich-magistrates-court-trial-a8655361.html; Mark Thomas, "I've Been a Silly Man, Haven't I," *Legal Eagle,* RSPB Investigations Newsletter, Spring 2019, 4, https://ww2.rspb.org.uk/Images/Legal%20Eagle%2087_tcm9-465877.pdf.

21. Peter Walker, "Rare Bird Egg Thief, with Collection of 700 Snatched from Nests, Jailed," *Guardian,* December 13, 2011, https://www.theguardian.com/uk/2011/dec/13/prolific-egg-thief-700-jailed; Rubinstein, "Operation Easter"; Thomas, interview.

22. Wheeler, *Poached*; Thomas, interview.

23. Thomas, interview.

24. Thomas, interview.

25. Thomas, interview.

CHAPTER 3: EGG RUSH

1. Ian Webster, "$36 in 1849 → 2022/Inflation Calculator," Official Inflation Data, Alioth Finance, March 28, 2022, https://www.officialdata.org/us/inflation/1849?amount=36; "History of the Hangtown Fry and Recipes," City of Placerville, California, accessed November 23, 2021, https://www.cityofplacerville.org/history-of-the-hangtown-fry-and-recipes; History.com Editors, "San Francisco," History Channel, last modified December 18, 2009, https://www.history.com/topics/us-states/san-francisco.

2. Zach Coffman, "100 Years of the Farallon National Wildlife Refuge," *Tideline* 30, no. 1 (Spring 2009), accessed November 24, 2021, https://www .fws.gov/uploadedFiles/Region_8/NWRS/Zone_2/San_Francisco_Bay_ Complex/tideline%20SPRING%2009C.pdf; Peter White, *The Farallon Islands: Sentinels of the Golden Gate* (San Francisco: Scottwell Associates, 1995), 7.

3. Susan Casey, *The Devil's Teeth: A True Story of Obsession and Survival among America's Great White Sharks* (New York: Henry Holt, 2005), 79.

4. White, *Farallon Islands*, 45–55.

5. Charles Nordhoff, "The Farallon Islands," *Harper's New Monthly Magazine* 48, no. 287 (April 1874): 617–625, https://catalog.hathitrust.org/ Record/000505748/Home.

6. "Aid Carried to Marooned Egg Hunters by the *Call*'s Stanch Tug *Reliance*," *San Francisco Call* 86, no. 12 (July 12, 1899).

7. White, *Farallon Islands*, 43; Casey, *Devil's Teeth*, 82.

8. "Nerva N. Wines," J. Candace Clifford Lighthouse Research Catalog, accessed November 23, 2021, https://archives.uslhs.org/people/nerva-n -wines; Amos Clift letter quoted in White, *Farallon Islands*, 43.

9. White, *Farallon Islands*, 52–53.

10. White, *Farallon Islands*, 53–54.

11. White, *Farallon Islands*, 54.

12. Peter Pyle, "Seabirds," US Geological Survey Publications Warehouse, accessed March 28, 2022, https://pubs.usgs.gov/circ/c1198/chapters/150 -161_Seabirds.pdf; William J. Sydeman, "Survivorship of Common Murres on Southeast Farallon Island, California," *Ornis Scandinavica (Scandinavian Journal of Ornithology)* 24, no. 2 (1993): 135–141, https://doi.org/10.2307/3676363; "Important Bird Areas: Farallon Islands," Audubon, last modified May 10, 2018, https://www.audubon .org/important-bird-areas/farallon-islands.

13. Quoted in Errol Fuller, *The Great Auk: The Extinction of the Original Penguin* (Piermont, NH: Bunker Hill, 2003), 34.

14. Fuller, *Great Auk*, 82–83.

CHAPTER 4: EGG MONEY

1. Amina is not her real name, which I've changed for her privacy.

2. Emelyn Rude, *Tastes Like Chicken: A History of America's Favorite Bird* (New York: Pegasus, 2016), 6, 18–22, 33.

3. Jessica B. Harris, *High on the Hog: A Culinary Journey from Africa to America* (New York: Bloomsbury, 2011), 83–84.

4. Rude, *Tastes Like Chicken*, 33.

5. Adrian Miller, "The Surprising Origin of Fried Chicken," BBC, October

13, 2020, https://www.bbc.com/travel/article/20201012-the-surprising -origin-of-fried-chicken; Psyche A. Williams-Forson, *Building Houses out of Chicken Legs: Black Women, Food, and Power* (Chapel Hill: University of North Carolina Press, 2006), 1, 30–36.

6. Elizabeth A. Payne, "Egg Money Shaped Farm Women's Economy," *Daily Journal*, July 24, 2006; Andrew Lawler, *Why Did the Chicken Cross the World? The Epic Saga of the Bird That Powers Civilization* (New York: Atria Paperback, 2014), 203–204; City of Mansfield, MO, "Where the Little House Books Were Written," accessed January 2, 2022, http:// mansfieldcityhall.org/info.html; Vivana A. Zelizer, *The Social Meaning of Money: Pin Money, Paychecks, Poor Relief, and Other Currencies* (Princeton, NJ: Princeton University Press, 2017), 42, 62, 222.

7. Laura Ingalls Wilder, *Laura Ingalls Wilder, Farm Journalist: Writings from the Ozarks*, ed. Stephen W. Hines (Columbia: University of Missouri Press, 2007), 48–50.

8. Vittoria Traverso, "The Egyptian Egg Ovens Considered More Wondrous Than the Pyramids," Atlas Obscura, March 29, 2019, https://www .atlasobscura.com/articles/egypt-egg-ovens.

9. Diane Toops, *Eggs: A Global History*, Edible Series (London: Reaktion, 2014), 83–84; Bill Hammerman, "Lest We Forget—Lyman Byce," *Petaluma Argus-Courier*, July 17, 2014, https://bill-hammerman.blogs .petaluma360.com/13099/lest-we-forget-lyman-byce/; George Pendle, "The California Town That Produced 10 Million Eggs a Year," Atlas Obscura, July 25, 2016, https://www.atlasobscura.com/articles/the -california-town-that-produced-10-million-eggs-a-year; Dan Strehl, "Egg Basket of the World," in *Eggs in Cookery: Proceedings of the Oxford Symposium on Food and Cookery 2006*, ed. Richard Hosking (Devon, UK: Prospect, 2007), 246; "Christopher Nisson Archives," Petaluma Historian, accessed January 6, 2022, https://petalumahistorian .com/tag/christopher-nisson/; Diane Peterson, "History of Petaluma Eggs," *Sonoma Magazine*, March 2015, https://www.sonomamag .com/history-petaluma-eggs/.

10. Strehl, "Egg Basket of the World," 247.

11. Strehl, "Egg Basket of the World," 248.

12. Rude, *Tastes Like Chicken*, 110, 118–119.

13. "Cal-Maine Foods' Leadership Team," Cal-Maine Foods, accessed January 6, 2022, https://www.calmainefoods.com/company/cal-maine -foods-leadership-team/.

14. 29 CFR § 780.328, Meaning of Livestock, amended April 1, 2022; Veronica Hirsch, "Detailed Discussion of Legal Protections of the Domestic Chicken in the United States and Europe/Animal Legal and Historical Center," accessed January 6, 2022, https://www.animallaw.info/article/detailed -discussion-legal-protections-domestic-chicken-united-states-and-europe.

15. Hirsch, "Detailed Discussion of Legal Protection"; Temple Grandin, "Animal Welfare and Society Concerns Finding the Missing Link," *Meat Science* 98, no. 3 (November 2014), 466, https://doi.org/10.1016/j.meatsci.2014.05.011.

16. A. Iqbal and A. F. Moss, "Review: Key Tweaks to the Chicken's Beak—the Versatile Use of the Beak by Avian Species and Potential Approaches for Improvements in Poultry Production," *Animal* 15, no. 2 (February 2021), https://www.sciencedirect.com/science/article/pii/S175173112030121X; H. Cheng, "Morphological Changes and Pain in Beak Trimmed Laying Hens," *World's Poultry Science Journal* 62, no. 1 (2006), https://doi.org/10.1079/WPS200583; Michael J. Gentle et al., "Behavioral Evidence for Persistent Pain Following Partial Beak Amputation in Chickens," *Applied Animal Behaviour Science* 27, no. 1–2 (August 1990): 149–157, accessed December 14, 2021, https://doi.org/10.1016/0168-1591(90)90014-5.

17. P. Y. Hester and H. Shea-Moore, "Beak Trimming Egg-Laying Strains of Chickens," *World's Poultry Science Journal* 59, no. 4 (2003), accessed December 14, 2021, https://doi.org/10.1079/WPS20030029.

18. Bill Gates, "Why I Would Raise Chickens," Gatesnotes.com, accessed January 6, 2022, https://www.gatesnotes.com/development/why-i-would-raise-chickens; Melinda Gates, "The Small Animal That's Making a Big Difference for Women in the Developing World," Medium.com, last modified June 10, 2016, https://medium.com/bill-melinda-gates-foundation/the-small-animal-thats-making-a-big-difference-for-women-in-the-developing-world-15d31dca2cc2.

CHAPTER 5: EGG GURUS

1. Jacques Pépin, phone interview, October 2, 2020; Auguste Escoffier, *A Guide to Modern Cookery* (London: W. Heinemann, 1907), 164; Harold McGee, *On Food and Cooking: The Science and Lore of the Kitchen* (New York: Scribner, 2004), 68–69.

2. J. Kenji López-Alt, phone interview, February 24, 2021.

3. General Mills, *Betty Crocker's Cookbook* (New York: Golden, 1974), 205.

4. "Eggs: The Perfect Balance of Yin and Yang," Traditional Chinese Medicine World Foundation, May 7, 2021, https://www.tcmworld.org/eggs-perfect-balance-yin-yang/.

5. McGee, *On Food*, 76.

6. McGee, *On Food*, 84–86.

7. Paul Freedman, phone interview, September 24, 2020.

8. Details about Jacques and his life throughout are drawn from Jacques

Pépin, *The Apprentice: My Life in the Kitchen* (Boston: Houghton Mifflin Harcourt, 2004); and Pépin, phone interviews, October 2, 2020, and March 17, 2021.

9. Rupert Taylor, "The Mysterious Origin of Eggs Benedict," Delishably, November 15, 2021, https://delishably.com/dairy/The-Mysterious-Origin-of-Eggs-Benedict; Pépin, phone interviews.

10. López-Alt, phone interview; Pierre Bourdieu, *Distinction: A Social Critique of the Judgement of Taste* (Cambridge, MA: Harvard University Press, 1984).

11. Pépin, phone interviews; Maguelonne Toussaint-Samat, *A History of Food* (West Sussex, UK: Blackwell, 2009), 326; Freedman, phone interview; Ken Albala, phone interview, September 22, 2020.

CHAPTER 6: VELVET EGGS

1. Gina L. Greco and Christine M. Rose, trans., *The Good Wife's Guide (Le Ménagier de Paris: A Medieval Household Book)* (Ithaca, NY: Cornell University Press, 2009), 310–311.

2. Mihaly Csikszentmihalyi, *Flow: The Psychology of Optimal Experience* (New York: HarperCollins, 2009).

3. Jacques Pépin, phone interviews, October 2, 2020, March 17, 2021, and February 8, 2022.

4. Pépin, phone interviews.

5. Lynne Rossetto Kasper, "There's More Than One Way to Cook an Egg: Dave Arnold Has 11," Splendid Table, April 12, 2013, https://www.splendidtable.org/story/2013/04/12/theres-more-than-one-way-to-cook-an-egg-dave-arnold-has-11.

6. Wei Guo, "Chinese Steamed Eggs, a Perfectionist's Guide," Red House Spice, last modified December 24, 2019, https://redhousespice.com/chinese-steamed-eggs/. See also one of my favorite Chinese chefs, Daddy Lau, "Steamed Egg (蒸蛋)," Chinese Family Recipes/Made with Lau, last modified August 20, 2020, https://madewithlau.com/recipes/steamed-egg.

CHAPTER 7: PYSANKY

1. Angela Hui, "Why My Childhood Birthdays Were Full of Red Eggs," Goldthread, February 6, 2019, https://www.goldthread2.com/food/why-my-childhood-birthdays-were-full-red-eggs/article/3000730; Theresa Vargas, "The Revered (and Very Messy) Easter Tradition You Might Not Have Heard About," *Washington Post*, April 20, 2019, https://

www.washingtonpost.com/local/the-revered-and-very-messy-easter
-tradition-you-might-not-have-heard-about/2019/04/19/c623d4e4
-62f0-11e9-9412-daf3d2e67c6d_story.html; "Egg-shoeing in Focus
for Easter as Hungarian Craftsman Keeps Tradition Alive," Reuters,
last modified March 31, 2021, https://www.tribuneindia.com/news/
schools/egg-shoeing-in-focus-for-easter-as-hungarian-craftsman
-keeps-tradition-alive-232717.

2. Brian Stewart, "Egg Cetera #6: Hunting for the World's Oldest Decorated Eggs," University of Cambridge, April 10, 2012, https://www
.cam.ac.uk/research/news/egg-cetera-6-hunting-for-the-worlds-oldest
-decorated-eggs; Jonathan Amos, "Etched Ostrich Eggs Illustrate
Human Sophistication," BBC News, last modified March 2, 2010,
https://news.bbc.co.uk/2/hi/science/nature/8544332.stm.

3. Stewart, "Egg Cetera #6"; John P. Rafferty, "6 of the World's Most Dangerous Birds," *Encyclopedia Britannica*, accessed January 20, 2022,
https://www.britannica.com/list/6-of-the-worlds-most-dangerous
-birds.

4. Malayna Evans Williams, *Signs of Creation: Sex, Gender, Categories,
Religion, and the Body in Ancient Egypt*, PhD diss., University of Chicago, June 2011, 146; Sara El Sayed Kitat, "Ostrich Egg and Its Symbolic Meaning in the Ancient Egyptian Monastery Churches," *Journal
of the General Union of Arab Archeologists* 15, no. 15 (Winter 2014): 25,
https://journals.ekb.eg/article_3088.html.

5. Tamar Hodos, "Eggstraordinary Artefacts: Decorated Ostrich Eggs in
the Ancient Mediterranean World," *Humanities and Social Sciences
Communications* 7, no. 1 (2020), doi:10.1057/s41599-020-00541-8; El
Sayed Kitat, "Ostrich Egg and Its Symbolic Meaning," 24; Francesco
Careilli, "The Book of Death: Weighing Your Heart," *London Journal of
Primary Care* 4, no. 1 (July 2011), doi: 10.1080/17571472.2011.11493336;
John Habib, "Do You Know These 4 Orthodox Church Symbols?"
Orthodox Christian Meets World blog, June 30, 2015, last modified
March 29, 2020, https://johnbelovedhabib.wordpress.com/2015/06/30/
do-you-know-these-4-orthodox-church-symbols/comment-page-1/;
Martin Kemp, "Science in Culture: Eggs and Exegesis," *Nature* 440, no.
7086 (2006): 872, doi:10.1038/440872a.

6. In addition, in Turkey's Blue Mosque, as well as the Hagia Sophia,
ostrich eggs are hung in light fixtures because they release an odor that
repels spiders, thus reducing unsightly cobwebs. Rabah Saoud, "Sultan
Ahmet Cami or Blue Mosque," Muslim Heritage, July 8, 2004, https://
muslimheritage.com/sultan-ahmet-cami-blue-mosque/; Mary D. Gar-
rard, *Brunelleschi's Egg: Nature, Art, and Gender in Rennaisance Italy*
(Los Angeles: University of California Press, 2010), 45; Creighton Gil-
bert, "'The Egg Reopened' Again," *Art Bulletin* 56, no. 2 (1974): 252–

258, https://doi.org/10.2307/3049230; Millard Meiss, "Not an Ostrich Egg?" *Art Bulletin* 57, no. 1 (1975): 116, doi:10.2307/3049344.

7. Venetia Newall, *An Egg at Easter: A Folklore Study* (London: Routledge & Kegan Paul, 1971), 268.

8. Newall, *An Egg at Easter,* 263–264.

9. Or so my Swedish friends inform me about Easter crones. Newall, *An Egg at Easter,* 124–125, 208; "Easter Festivities in Slovenia," I Feel Slovenia, accessed April 4, 2022, https://www.slovenia.info/en/stories/easter -in-slovenia; Gabriel Stille, "Śmigus-Dyngus: Poland's National Water Fight Day," Culture.ple, March 11, 2014, https://culture.pl/en/article/ smigus-dyngus-polands-national-water-fight-day.

10. Luba Petrusha, "Soviet Era," Pysanky.info, accessed April 4, 2022, https://www.pysanky.info/History/Soviet.html; Vira Manko, *The Ukrainian Folk Pysanka* (Lviv: Svichado, 2017), 11; "Pysanka Symbols and Motifs with Luba Petrusha," Ukrainian History and Education Center, accessed April 4, 2022, https://www.ukrhec.org/civicrm/event/ info%3Fid%3D128%26reset%3D1; Luba Petrusha, "Oleska Voropay," accessed April 5, 2021, https://pysanky.info/pysanka_legends/voropay .html; Theresa Vargas, "In Ukrainian Eggs, People Are Finding a Way to Connect and Help," *Washington Post,* March 30, 2022, https://www .washingtonpost.com/dc-md-va/2022/03/30/ukraine-eggs-pysanky -easter.

11. "Pysanky—Ukrainian Easter Eggs," Ukrainian Museum in New York City, last modified 2011, https://www.ukrainianmuseum.org/ ex_110326pysanka.html.

12. Yaroslava Tkachuk, "Pysanka: Easter Traditions," National Museum of Hutsul Region and Pokuttya, accessed January 20, 2022, https://pysanka .museum/museum/articles/easter_traditions/.

13. Luba Petrusha, "Ancient Origins," Pysanka, accessed January 20, 2022, https://www.pysanky.info/History/Ancient.html; Petrusha, "Kyivan Rus," Pysanka, https://www.pysanky.info/History/Kyivan_Rus.html; Meredith Bennett-Smith, "Look: 500-Year-Old Easter Egg?" HuffPost, August 9, 2013, last modified December 7, 2017, https://www.huffpost .com/entry/easter-egg-ukraine-500-year-old-photo_n_3732610.

14. Marian J. Rubchak, "Ukraine's Ancient Matriarch as a Topos in Constructing a Feminine Identity," *Feminist Review* 92, no. 1 (2009): 131, 133, doi:10.1057/fr.2009.5.

15. Arricca Elin Sansone, "How Did Colorful Decorated Eggs Become a Symbol of Easter?" *Country Living,* last modified March 20, 2019, https://www.countryliving.com/life/a26388851/history-of-easter-eggs/; "Pysanky—Ukrainian Easter Eggs—Ukrainian Museum (NYC) Exhibits/Lectures," Ukrainian Museum, New York City, accessed January 20, 2022, https://www.ukrainianmuseum.org/ex_100306pysanka.html.

16. "The Imperial Eggs," Fabergé.com, accessed January 20, 2022, https://www.faberge.com/the-world-of-faberge/the-imperial-eggs.

17. Miss Justina Marie, "Pysanky 101: How to Make Ukrainian Easter Eggs (Tutorial for Beginners) + Tour of My Pysanky," YouTube, March 28, 2019, https://www.youtube.com/watch?v=LjcKizt9n5A.

18. There are more colors of pysanky dye than those named in the text, but my little starter pack came with only six basic colors. For more info, see Luba Petrusha, "Color Sequences," Pysanka, accessed January 20, 2022, https://www.pysanky.info/Dyeing/Dye_Sequences.html.

19. Electric *kistkas* are also available for the true devotee of the art, and that is what Miss Justina Marie prefers.

20. Miss Justina Marie, "Pysanky 101."

CHAPTER 8: CLOWN EGGS

1. Pieter Aertsen, *The Egg Dance*, oil on panel, 1552 (Rijksmuseum, Amsterdam), https://www.rijksmuseum.nl/en/rijksstudio/artists/pieter-aertsen/objects#/SK-A-3,0; Jan Steen, *The Egg Dance: Peasants Merrymaking in an Inn*, oil on canvas, 1670s (Wellington Collection, London), https://artuk.org/discover/artworks/the-egg-dance-peasants-merrymaking-in-an-inn-144403.

2. Mattie Faint, phone interview, May 4, 2021.

3. Luke Stephenson and Helen Champion, *The Clown Egg Register* (San Francisco: Chronicle, 2018), 216.

4. Stephenson and Champion, *Clown Egg Register*, 215–216; Faint, interview.

5. Faint, interview.

6. BeWell@Stanford, "The Opposite of Play Is Not Work—It Is Depression," Wu Tsai Neurosciences Institute, Stanford University, last modified May 29, 2015, https://neuroscience.stanford.edu/news/opposite-play-not-work-it-depression. Much of what I say about play and the magic circle here represents my own understanding as a designer, researcher, and partcipant, informed by rigorous scholarly debate around, for example, whether the "magic circle" truly exists and, if so, how porous it is. To read more, see Jaakko Stenros, *Playfulness, Play, and Games: A Constructionist Ludology Approach*, PhD diss., University of Tampere, Finland, 2015; Stenros, "In Defence of a Magic Circle: The Social, Mental and Cultural Boundaries of Play," DiGRA Nordic 2012 Conference: Local and Global—Games in Culture and Society, Tampere, Finland, June 6–8, 2012, edited by Raine Koskimaa, Frans Mäyrä, and Jaakko Suominen; Markus Montola, "The Positive Negative Experience in Extreme Role-Playing," *Proceedings of DiGRA Nordic 2010: Experi-*

encing Games—Games, Play, and Players, Stockholm, Sweden, August 16, 2010; Montola, *On the Edge of the Magic Circle: Understanding Pervasive Games and Role-Playing*, PhD diss., University of Tampere, Finland, 2012; Sarah Lynne Bowman and Kjell Hedgard Hugaas, "Magic Is Real: How Role-Playing Can Transform Our Identities, Our Communities, and Our Lives," in *Book of Magic: Vibrant Fragments of Larp Practices*, ed. Kari Kvittingen Djukastein, Marcus Irgens, Nadja Lipsyc, and Lars Kristian Løveng Sunde (Oslo: Knutepunkt, 2021), 52–74.

7. Debbie Smith, phone interview, May 13, 2021.

8. Smith, interview.

9. Smith, interview.

10. David Fagundes and Aaron Perzanowski, "Clown Eggs," *University of Notre Dame Australia Law Review* 94, no. 3 (n.d.): 1313–1380, doi:10.32613/undalr/2017.19; Smith, interview.

11. Smith, interview.

CHAPTER 9: EGG TOSS

1. "Food Fight: Festival in Spain Holds a Flour-and-Egg Battle," AP News, December 28, 2018, https://apnews.com/article/entertainment-lifestyle-travel-spain-els-enfarinats-7b6119007c6949108be5024634b01352; Matthew White, "Crime and Punishment in Georgian Britain," British Library, October 14, 2009, https://www.bl.uk/georgian-britain/articles/crime-and-punishment-in-georgian-britain; A. R. T. Kemasang, "The Egg in European Diet and What It Tells Us," *Petits Propos Culinaires* 115 (October 2019): 91; Samuel Osborne, "Over 2,300 People Pledge to Take Part in Egg-Throwing Contest at Margaret Thatcher Statue Unveiling," *Independent*, December 1, 2020, https://www.independent.co.uk/news/uk/politics/margaret-thatcher-statue-grantham-egg-throwing-contest-b1764620.html; Cameron Wilson, "A Boy Egged a Racist Politician after Christchurch: A Year On, Their Lives Have Completely Changed," BuzzFeed, March 15, 2020, https://www.buzzfeed.com/cameronwilson/will-connolly-fraser-anning-christchurch-attack-egg-boy.

2. "British Crowd Hurls Eggs at Nazi Leader and Bride," *New York Times*, October 6, 1963, 3, https://nyti.ms/3K4bjli; Joanne Kavanagh, "Who Was Colin Jordan's Wife Françoise Dior?" *US Sun*, September 28, 2021, https://www.the-sun.com/news/3751183/colin-jordan-wife-francoise-dior/.

3. Karen Chernick, "How Eggs Became an Unlikely but Popular Material for Painters and Photographers," Artsy, March 30, 2018, https://www.artsy.net/article/artsy-editorial-eggs-popular-material-painters-photographers.

4. Arie Wallert, "*Libro secondo de diversi colori e sise da mettere a oro*: A Fifteenth-Century Technical Treatise on Manuscript Illumination," in *Historical Painting Techniques, Materials, and Studio Practice: Preprints of a Symposium, University of Leiden, the Netherlands, 26–29 June 1995* (Los Angeles: Getty Publications, 1995), 4–78.

5. Carolin C. Young, "Salvador Dali's Giant Egg," in *Eggs in Cookery: Proceedings of the Oxford Symposium of Food and Cookery 2006*, ed. Richard Hosking (Devon, UK: Prospect, 2007), 293–294, 302.

6. Hannah Yi, "This French Guy Is Sitting on Eggs until They Hatch: It's Art," Quartz, March 31, 2017, https://qz.com/946468/this-french-guy-is -sitting-on-eggs-until-they-hatch-its-art/; Agence France-Presse (AFP), "'Human Hen' Artist Condemned after Hatching Nine Eggs," NDTV. com, last modified April 21, 2017, https://www.ndtv.com/offbeat/human -hen-artist-condemned-after-hatching-nine-eggs-1684057.

7. Milo Moiré, "The 'PlopEgg' Painting Performance #1 (Art Cologne 2014)," YouTube, April 12, 2014, https://www.youtube.com/ watch?v=wKFZOIv5sS0.

8. "I Paint with My Prick," Quote Investigator—Tracing Quotations, May 28, 2012, https://quoteinvestigator.com/2012/05/28/renoir-paint/.

9. The "mental load" has been around in feminist spaces for a while. I first encountered it here: Emma Clit, "You Should Have Asked," May 20, 2017, last accessed November 30, 2021, https://english.emmaclit .com/2017/05/20/you-shouldve-asked/.

10. "Young British Artists (YBAs)—Art Term," Tate, accessed February 28, 2022, https://www.tate.org.uk/art/art-terms/y/young-british-artists -ybas; Zsofia Paulikovics, "A Guide to the Controversial Works of YBA Sarah Lucas," Dazed Digital, last modified October 8, 2018, https:// www.dazeddigital.com/art-photography/article/41690/1/guide-to -controversial-yba-sarah-lucas-au-naturel-new-museum-exhibition.

11. Sarah Vankin, "Eggs Are Being Fried as Art at the Hammer Museum: Let's See What's Cooking," *Los Angeles Times*, July 9, 2019, https:// www.latimes.com/entertainment/arts/la-et-cm-sarah-lucas-cooking -eggs-20190709-story.html; Ellie Howard, "British Pavilion: Who Is Sarah Lucas?" Kids of Dada, accessed February 28, 2022, https://www .kidsofdada.com/blogs/magazine/19753281-british-pavilion-who-is -sarah-lucas.

12. Sarah Lucas and Julian Simmons, "Egg Massage," *Male Nudes: A Salon from 1800 to 2021*, 2015, https://website-artlogicwebsite0032 .artlogic.net/viewing-room/34-male-nudes-a-salon-from-1800-to -2021/; Zachary Small, "Sarah Lucas Makes Male Privilege Her Own," Hyperallergic, last modified November 2, 2018, https://hyperallergic .com/463286/sarah-lucas-makes-male-privilege-her-own/.

13. Digby Warde-Aldam, "The Shock Factor of Sarah Lucas," *Apollo Mag-*

azine, last modified November 22, 2018, https://www.apollo-magazine.com/sarah-lucas-shock-value/.

14. Lorissa W. Rinehart, "Why Splattering Eggs on a Museum's Walls with Other Women Was So Satisfying," Hyperallergic, May 28, 2019, https://hyperallergic.com/502398/why-splattering-eggs-on-a-museums-walls-with-other-women-was-so-satisfying/.

15. Aida Edemariam, "The Saturday Interview: Sarah Lucas," *Guardian*, last modified May 27, 2011, https://www.theguardian.com/theguardian/2011/may/27/the-saturday-interview-sarah-lucas; Christina Patterson, "Sarah Lucas: A Young British Artist Grows Up and Speaks Out," *Independent*, last modified July 20, 2012, https://www.independent.co.uk/arts-entertainment/art/features/sarah-lucas-a-young-british-artist-grows-up-and-speaks-out-7959882.html; Hannah Alberico et al., "Workflow Optimization for Identification of Female Germline or Oogonial Stem Cells in Human Ovarian Cortex Using Single-Cell RNA Sequence Analysis," *Stem Cells*, March 9, 2022, https://doi.org/10.1093/stmcls/sxac015.

CHAPTER 10: SPACE EGGS

1. Dorothy Parker and Stuart Y. Silverstein, *Not Much Fun: The Lost Poems of Dorothy Parker* (New York: Scribner, 2009), 33; Marion Meade, *Dorothy Parker: What Fresh Hell Is This?* (New York: Villard, 1989), 105.

2. Katie Valentine, "The Amazing Story of the Cold War Space-Egg Race," Audubon, December 15, 2017, https://www.audubon.org/news/the-amazing-story-cold-war-space-egg-race; SaVanna Shoemaker, "Quail Eggs: Nutrition, Benefits, and Precautions," Healthline, February 24, 2020, https://www.healthline.com/nutrition/quail-eggs-benefits#comparison-with-chicken-eggs.

3. "Victor F. Hess—Facts," NobelPrize.org, accessed February 28, 2022, https://www.nobelprize.org/prizes/physics/1936/hess/facts/.

4. Jordan Bimm, email correspondence, January 6, 2020.

5. Dietrich E. Beischer and Alfred R. Fregly, *Animals and Man in Space: A Chronology and Annotated Bibliography through the Year 1960* (Pensacola, FL: Office of Naval Research, Department of the Navy, 1962), 27.

6. JPat Brown, "Cooking with FOIA: The CIA's Top Secret Anti-Poop Diet," MuckRock, last modified October 16, 2018, https://www.muckrock.com/news/archives/2018/oct/16/cooking-foia-air-forces-top-secret-u-2-pilot-diet/.

7. European Space Agency, "Space Scrambled Eggs," YouTube, January 30, 2016, https://www.youtube.com/watch?v=MtNGI-tFZxU.

8. John Vellinger and Mark Deuser, phone interview, November 23, 2020.

9. "John Vellinger: From Chix in Space to a Company in Space," Mechanical Engineering, Purdue University, accessed March 1, 2022, https://engineering.purdue.edu/ME/News/john-vellinger-from-chix-in-space-to-a-company-in-space; Vellinger and Deuser, interview.

10. Vanessa Listek, "Techshot's Bioprinter Successfully Fabricates Human Menisci in Space," 3DPrint.com, last modified October 16, 2021, https://3dprint.com/265654/techshots-bioprinter-successfully-fabricated-human-menisci-in-space/; "Redwire Acquires Techshot, the Leader in Space Biotechnology," Redwire Space, last modified November 2, 2021, https://redwirespace.com/newsroom/redwire-acquires-techshot-the-leader-in-space-biotechnology/.

11. Stephen Doty, phone interview, February 9, 2021.

12. Doty, interview.

13. Rich Boling, phone interview, February 16, 2022; Doty, interview; Ann Hutchinson, "Quail Eggs to Provide Clues to Effects of Microgravity," NASA, November 26, 2001, https://www.nasa.gov/centers/ames/news/releases/2001/01_91AR.html; "In the Heartland of Hearing Research: Central Institute for the Deaf," Hearing Review, last modified November 1, 2001, https://hearingreview.com/inside-hearing/research/in-the-heartland-of-hearing-research-central-institute-for-the-deaf.

14. Boling, interview.

CHAPTER 11: EGG CURES

1. Jessie Yeung, "The US Keeps Millions of Chickens in Secret Farms to Make Flu Vaccines: But Their Eggs Won't Work for Coronavirus," CNN, March 29, 2020, https://www.cnn.com/2020/03/27/health/chicken-egg-flu-vaccine-intl-hnk-scli/index.html; James Pasley, "The US Government Has Possibly Millions of Chickens in Secret Locations Laying Eggs Year Round for Flu Vaccines: The Exact Number and Location Are a Matter of National Security—Here's What We Know about the Chickens," Insider, April 7, 2020, https://www.insider.com/us-government-flu-vaccine-chickens-national-security-2020-4.

2. "Smallpox," Cleveland Clinic, accessed March 1, 2022, https://my.clevelandclinic.org/health/diseases/10855-smallpox.

3. History of Vaccines—"How Are Vaccines Made? The Scientific Method in Vaccine History," accessed June 14, 2022, https://historyofvaccines.org/vaccines-101/how-are-vaccines-made/scientific-method-vaccine-history; Steven Johnson, "How Humanity Gave Itself an Extra Life,"

New York Times Magazine, April 27, 2021, https://www.nytimes.com/2021/04/27/magazine/global-life-span.html.

4. Sam Kean, "22 Orphans Gave Up Everything to Distribute the World's First Vaccine," *Atlantic*, January 12, 2021, https://www.theatlantic.com/science/archive/2021/01/orphans-smallpox-vaccine-distribution/617646/.

5. History of Vaccines—"How Are Vaccines Made? The Scientific Method in Vaccine History," accessed June 14, 2022, https://historyofvaccines.org/vaccines-101/how-are-vaccines-made/scientific-method-vaccine-history.

6. Centers for Disease Control and Prevention, National Center for Immunization and Respiratory Diseases (NCIRD), "History of 1918 Flu Pandemic," Centers for Disease Control and Prevention, last modified January 22, 2019, https://www.cdc.gov/flu/pandemic-resources/1918-commemoration/1918-pandemic-history.htm; Christopher Ryland, curator, History of Medicine Collections and Archives at Vanderbilt University Libraries, personal correspondence, November 2, 2021; Greer Williams, *Virus Hunters* (London: Hutchinson, 1959), 109.

7. Williams, *Virus Hunters*, 100, 114–115, 136; Esmond R. Long, "Ernest William Goodpasture: 1886–1960, A Biographical Memoir," National Academy of Sciences, last modified 1965, https://www.nasonline.org/publications/biographical-memoirs/memoir-pdfs/goodpasture-ernest.pdf; Alice M. Woodruff and Ernest W. Goodpasture, "The Susceptibility of the Chorio-Allantoic Membrane of Chick Embryos to Infection with the Fowl-Pox Virus," *American Journal of Pathology* 7, no. 3 (May 1931): 209–222.5, https://www.ncbi.nlm.nih.gov/pmc/articles/PMC2062632/.

8. Williams, *Virus Hunters*, 110, 114–116, 136; Woodruff and Goodpasture, "Susceptibility of Chorio-Allantoic Membrane"; Ernest W. Goodpasture, Alice M. Woodruff, and Gerrit J. Buddingh, "Vaccinal Infection of the Chorio-Allantoic Membrane of the Chick Embryo," *American Journal of Pathology* 8, no. 3 (May 1932): 271–282.7, https://www.ncbi.nlm.nih.gov/pmc/articles/PMC2062681/.

9. Williams, *Virus Hunters*, 109, 116; Leonard Norkin, "Tag Archives: Alice Woodruff," Leonard Norkin Virology Site, last modified December 10, 2014, https://norkinvirology.wordpress.com/tag/alice-woodruff/; Woodruff and Goodpasture, "Susceptibility of Chorio-Allantoic Membrane."

10. Long, "Goodpasture," 131–143; Ernest W. Goodpasture and Katherine Anderson, "Virus Infection of Human Fetal Membranes Grafted on the Chorioallantois of Chick Embryos," *American Journal of Pathology* 18, no. 4 (July 1942): 563–575; Ernest W. Goodpasture and Katherine Anderson, "Infection of Human Skin, Grafted on the Chorioallantois

of Chick Embryos, with the Virus of Herpes Zoster," *American Journal of Pathology* 30, no. 3 (May 1944): 447–455; Katherine Anderson, "Infection of Newborn Syrian Hamsters with the Virus of Mare Abortion," *American Journal of Pathology* 18 (July 1942): 555–561; "A Chicken's Egg (1931)," British Society for Immunology, accessed March 2, 2022, https://www.immunology.org/chickens-egg-1931; "Influenza Historic Timeline," Centers for Disease Control and Prevention, last modified April 18, 2019, https://www.cdc.gov/flu/pandemic-resources/pandemic-timeline-1930-and-beyond.htm; Leigh Krietsch Boerner, "The Flu Shot and the Egg," *ACS Central Science* 6, no. 2 (February 2020): 89–92, doi:10.1021/acscentsci.0c00107; Eric Durr, "Worldwide Flu Outbreak Killed 45,000 American Soldiers during World War I," US Army (website), last modified August 31, 2018, https://www.army.mil/article/210420/worldwide_flu_outbreak_killed_45000_a.

11. Eric Bender, "Accelerating Flu Protection," *Nature* 573, no. 7774 (September 18, 2019), doi:10.1038/d41586-019-02756-5; David Robson, "The Real Reason Germs Spread in the Winter," BBC, October 18, 2015, https://www.bbc.com/future/article/20151016-the-real-reason-germs-spread-in-the-winter; "Selecting Viruses for the Seasonal Flu Vaccine," Centers for Disease Control and Prevention, last modified August 31, 2021, https://www.cdc.gov/flu/prevent/vaccine-selection.htm; Hien H. Nguyen, "Influenza Treatment and Management: Approach Considerations, Prevention, Prehospital Care," Medscape Reference, last modified November 5, 2021, https://emedicine.medscape.com/article/219557-treatment; Nancy Averett and Tania Elliott, "The Flu Vaccine Isn't 100% Effective, but Experts Recommend You Still Get It Every Year," Insider, May 17, 2021, https://www.insider.com/flu-vaccine-effectiveness; "Vaccine Effectiveness: How Well Do the Flu Vaccines Work?" Centers for Disease Control and Prevention, National Center for Immunization and Respiratory Diseases, last modified May 6, 2021, https://www.cdc.gov/flu/vaccines-work/vaccineeffect.htm.

12. "Human Cell Strains in Vaccine Development," History of Vaccines, College of Physicians of Philadelphia, last modified April 18, 2022, https://historyofvaccines.org/vaccines-101/how-are-vaccines-made/human-cell-strains-vaccine-development; Nicholas C. Wu et al., "Preventing an Antigenically Disruptive Mutation in Egg-Based H3N2 Seasonal Influenza Vaccines by Mutational Incompatibility," *Cell Host and Microbe* 25, no. 6 (June 2019), doi:10.1016/j.chom.2019.04.013; Krietsch Boerner, "Flu Shot and the Egg," 89–92; Elizabeth Pratt, "Why Do We Still Grow Flu Vaccines in Chicken Eggs?" Healthline, December 4, 2017, https://www.healthline.com/health-news/why-we-grow-flu-vaccines-in-chicken-eggs.

13. Please note that there are multiple types of vaccines. Because describing all of them is beyond the scope of this work, I give a high-level gloss

in the text. Some types include inactivated vaccines, which use dead viruses; live attenuated vaccines, which use a weakened version of a virus; subunit, recombinant, polysaccharide, and conjugate vaccines, which use chopped-up germ bits; toxoid vaccines, which contain inactivated toxins; and even more. "Different Types of Vaccines," History of Vaccines, College of Physicians of Philadelphia, 2018, accessed March 2, 2022, https://www.historyofvaccines.org/content/articles/different -types-vaccines; "Gene Synthesis Cost," Synbio Technologies: A DNA Technology Company, accessed March 2, 2022, https://www.synbio -tech.com/gene-synthesis-cost/; Elie Dolgin, "How COVID Unlocked the Power of RNA Vaccines," *Nature* 589, no. 7841 (2021): 189–191, doi:10.1038/d41586-021-00019-w.

14. The way B-cells function is even more complex than described here, but a full rendering of their function lies beyond the scope of this work. Dr. Biology, "B-Cells: Ask a Biologist," Arizona State University School of Life Sciences—Ask a Biologist, last modified February 16, 2011, https:// askabiologist.asu.edu/b-cell; "What Are Naïve Cells? Naïve T Cell, Naïve B Cell, and How to Isolate Naïve Lymphocytes," Akadeum Life Sciences, last modified April 14, 2021, https://www.akadeum.com/ blog/what-are-naive-cells/; Khan Academy, "B Lymphocytes (B Cells)/ Immune System Physiology," YouTube, February 18, 2010, https://www .youtube.com/watch?v=Z36dUduOk1Y.

15. Keith S. Kaye, "Comorbidities, Metabolic Changes Make Elderly More Susceptible to Infection," Healio: Medical News, Journals, and Free CME, September 1, 2011, https://www.healio.com/news/infectious -disease/20120225/comorbidities-metabolic-changes-make-elderly -more-susceptible-to-infection; Jonathan Abraham, "Passive Antibody Therapy in COVID-19," *Nature Reviews Immunology* 20, no. 7 (2020): 401–403, doi:10.1038/s41577-020-0365-7; "Passive Immunization," History of Vaccines, College of Physicians of Philadelphia, last modified April 11, 2022, https://www.historyofvaccines.org/index.php/content/ articles/passive-immunization.

16. Lucia Lee et al., "Immunoglobulin Y for Potential Diagnostic and Therapeutic Applications in Infectious Diseases," *Frontiers in Immunology* 12 (June 9, 2021), doi:10.3389/fimmu.2021.696003; Xiangguang Li et al., "Production and Characteristics of a Novel Chicken Egg Yolk Antibody (IgY) against Periodontitis-Associated Pathogens," *Journal of Oral Microbiology* 12, no. 1 (October 2020): 1831374, doi:10.1080/2000229 7.2020.1831374; Miriele C. Da Silva et al., "Production and Application of Anti-Nucleoprotein IgY Antibodies for Influenza A Virus Detection in Swine," *Journal of Immunological Methods* 461 (October 2018): 100–105, doi:10.1016/j.jim.2018.06.023; Yang Zhu et al., "Efficient Production of Human Norovirus-Specific IgY in Egg Yolks by Vaccination of Hens

with a Recombinant Vesicular Stomatitis Virus Expressing VP1 Protein," *Viruses* 11, no. 5 (May 2019): 444, doi:10.3390/v11050444; José M. Pérez de la Lastra et al., "Can Immunization of Hens Provide Oral-Based Therapeutics against COVID-19?" *Vaccines* 8, no. 3 (August 2020): 486, doi:10.3390/vaccines8030486; Mary Romeo, "A Neat Trick—Passive Immunization Using Chicken Antibodies," SPARK at Stanford, January 22, 2021, https://sparkmed.stanford.edu/blog/passive-immunization-using-chicken-antibodies; Sarah Graham, "Chicken Eggs Made to Produce Human Antibodies," *Scientific American*, August 29, 2005, https://www.scientificamerican.com/article/chicken-eggs-made-to-prod/.

17. Mark Cartwright, "Ancient Greek Medicine," in *World History Encyclopedia* (2018), accessed March 2, 2022, https://www.worldhistory.org/Greek_Medicine/; Yvette Brazier, "Ancient Roman Medicine: Influences, Practice, and Learning," Medical News Today, last modified November 9, 2018, https://www.medicalnewstoday.com/articles/323600#learning-about-the-body; Tram T. Vuong et al., "Processed Eggshell Membrane Powder Regulates Cellular Functions and Increase MMP-Activity Important in Early Wound Healing Processes," *PLOS One* 13, no. 8 (August 2018), doi:10.1371/journal.pone.0201975; Rosemond A. Mensah et al., "The Eggshell Membrane: A Potential Biomaterial for Corneal Wound Healing," *Journal of Biomaterials Applications* 36, no. 5 (November 2021): 912–929, doi:10.1177/08853282211024040; Susan Hewlings, Douglas Kalman, and Luke V. Schneider, "A Randomized, Double-Blind, Placebo-Controlled, Prospective Clinical Trial Evaluating Water-Soluble Chicken Eggshell Membrane for Improvement in Joint Health in Adults with Knee Osteoarthritis," *Journal of Medicinal Food* 22, no. 9 (September 2019): 875–884, doi:10.1089/jmf.2019.0068; Douglas S. Kalman and Susan Hewlings, "The Effect of Oral Hydrolyzed Eggshell Membrane on the Appearance of Hair, Skin, and Nails in Healthy Middle-Aged Adults: A Randomized Double-Blind Placebo-Controlled Clinical Trial," *Journal of Cosmetic Dermatology* 19, no. 6 (January 2020): 1463–1472, doi:10.1111/jocd.13275; Matej Baláž et al., "State-of-the-Art of Eggshell Waste in Materials Science: Recent Advances in Catalysis, Pharmaceutical Applications, and Mechanochemistry," *Frontiers in Bioengineering and Biotechnology* 8 (January 27, 2021), doi:10.3389/fbioe.2020.612567.

18. Rob Stein, "A Gene-Editing Experiment Let These Patients with Vision Loss See Color Again," NPR.org, September 29, 2021, https://www.npr.org/sections/health-shots/2021/09/29/1040879179/vision-loss-crispr-treatment; Michael Tabb, Andrea Gawrylewski, and Jeffery DelViscio, "What Is CRISPR, and Why Is It So Important?" *Scientific American*, video, June 22, 2021, https://www.scientificamerican.com/

video/what-is-crispr-and-why-is-it-so-important/; Rob Stein, "Blind Patients Hope Landmark Gene-Editing Experiment Will Restore Their Vision," NPR.org, May 10, 2021, https://www.npr.org/sections/health -shots/2021/05/10/993656603/blind-patients-hope-landmark-gene -editing-experiment-will-restore-their-vision.

19. Marek Bednarczyk et al., "Generation of Transgenic Chickens by the Non-Viral, Cell-Based Method: Effectiveness of Some Elements of This Strategy," *Journal of Applied Genetics* 59, no. 1 (January 2018): 81–89, doi:10.1007/s13353-018-0429-6; Ken-ichi Nishijima and Shinji Iijima, "Transgenic Chickens," *Development, Growth and Differentiation* 55, no. 1 (December 2012): 207–216, doi:10.1111/dgd.12032; Lei Zhu et al., "Production of Human Monoclonal Antibody in Eggs of Chimeric Chickens," *Nature Biotechnology* 23 (August 2005): 1159–1169, https:// www.nature.com/articles/nbt1132; Rachel Becker, "US Government Approves Transgenic Chicken," *Nature*, December 9, 2015, doi:10.1038/ nature.2015.18985; "The Only FDA-Approved Treatment for Lysosomal Acid Lipase Deficiency (LAL-D)," Alexion, AstraZeneca Rare Disease, accessed March 3, 2022, https://kanuma.com.

20. Takehiro Mukae et al., "Production of Recombinant Monoclonal Antibodies in the Egg White of Gene-Targeted Transgenic Chickens," *Genes* 12, no. 1 (January 2020): 38, doi:10.3390/genes12010038; Young M. Kim et al., "The Transgenic Chicken Derived Anti-CD20 Monoclonal Antibodies Exhibits Greater Anti-Cancer Therapeutic Potential with Enhanced Fc Effector Functions," *Biomaterials* 167 (June 2018): 58–68, doi:10.1016/j.biomaterials.2018.03.021; University of Edinburgh, "Hens That Lay Human Proteins in Eggs Offer Future Therapy Hope," EurekAlert!, January 27, 2019, https://www.eurekalert.org/ news-releases/607968; Kenneth Macdonald, "Gene Modified Chickens 'Lay Medicines,'" BBC News, January 28, 2019, https://www.bbc.com/ news/uk-scotland-47022070; Julie Kelly, "The March of Genetic Food Progress," *Wall Street Journal*, December 29, 2015, https://www.wsj .com/articles/the-march-of-genetic-food-progress-1451430187; Karthik Giridhar, "What to Know about HER2-Positive Breast Cancer," Mayo Clinic, last modified April 7, 2020, https://www.mayoclinic.org/breast -cancer/expert-answers/faq-20058066.

CHAPTER 12: HUMAN EGGS

1. Katherine Assersohn, Brekke Patricia, and Nicola Hemmings, "Physiological Factors Influencing Female Fertility in Birds," *Royal Society Open Science* 8 (2021): 2202274, http://doi.org/10.1098/rsos.202274; Adam M. Hawkridge, "The Chicken Model of Spontaneous Ovarian

Cancer," *Proteomics: Clinical Applications* 8, no. 9–10 (October 2014): 689–699, doi:10.1002/prca.201300135; M. F. Fathalla, "Incessant Ovulation and Ovarian Cancer: A Hypothesis Re-visited," *Fact, Views and Vision ObGyn* 5, no. 4 (2013): 292–297, https://www.ncbi.nlm.nih.gov/pmc/articles/PMC3987381/; "What Is Ovarian Cancer: Ovarian Tumors and Cysts," American Cancer Society, Information and Resources for Cancer: Breast, Colon, Lung, Prostate, Skin, accessed March 3, 2022, https://www.cancer.org/cancer/ovarian-cancer/about/what-is-ovarian -cancer.html; Tova M. Bergsten, Joanna E. Burdette, and Matthew Dean, "Fallopian Tube Initiation of High Grade Serous Ovarian Cancer and Ovarian Metastasis: Mechanisms and Therapeutic Implications," *Cancer Letters* 476 (April 2020): 152–160, doi:10.1016/j.canlet.2020.02.017; Hannah M. Micek et al., "The Many Microenvironments of Ovarian Cancer," *Advances in Experimental Medicine and Biology* 1296 (2020): 199–213, doi:10.1007/978-3-030-59038-3_12; Yang Yang-Hartwich et al., "Ovulation and Extra-Ovarian Origin of Ovarian Cancer," *Scientific Reports* 4, no. 1 (August 19, 2014), doi:10.1038/srep06116.

2.   My understanding of how cancer functions, and the car analogy in particular, derives from Siddhartha Mukherjee, *The Emperor of All Maladies: A Biography of Cancer* (New York: Simon & Schuster, 2010), 369.

3.   "Ovarian Cancer Risk Factors," American Cancer Society, Information and Resources for Cancer: Breast, Colon, Lung, Prostate, Skin, accessed March 3, 2022, https://www.cancer.org/cancer/ovarian-cancer/causes -risks-prevention/risk-factors.html.

4.   "Normal Ovarian Function," Rogel Cancer Center, University of Michigan, Ann Arbor, last modified April 10, 2018, https://www .rogelcancercenter.org/fertility-preservation/for-female-patients/normal -ovarian-function; Mariana Chavez-MacGregor et al., "Lifetime Cumulative Number of Menstrual Cycles and Serum Sex Hormone Levels in Postmenopausal Women," *Breast Cancer Research and Treatment* 108, no. 1 (March 2008): 101–112, doi:10.1007/s10549-007-9574-z; "Ovarian Cancer Statistics: How Common Is Ovarian Cancer," American Cancer Society, accessed March 3, 2022, https://www.cancer.org/cancer/ ovarian-cancer/about/key-statistics.html; Fathalla, "Incessant Ovulation"; Adam M. Hawkridge, "The Chicken Model of Spontaneous Ovarian Cancer."

5.   Patricia A. Johnson and James R. Giles, "The Hen as a Model of Ovarian Cancer," *Nature Reviews Cancer* 13, no. 6 (June 2013): 432–436, doi:10.1038/nrc3535; Erfan Eilati, Janice M. Bahr, and Dale B. Hales, "Long Term Consumption of Flaxseed Enriched Diet Decreased Ovarian Cancer Incidence and Prostaglandin E2 in Hens," *Gynecologic Oncology* 130, no. 3 (September 2013): 620–628, doi:10.1016/j .ygyno.2013.05.018; Southern Illinois University, "Flaxseed as Main-

tenance Therapy for Ovarian Cancer Patients in Remission," Clinical-Trials.gov, accessed March 3, 2022, https://clinicaltrials.gov/ct2/show/NCT02324439; Lindsey S. Treviño, Elizabeth L. Buckles, and Patricia A. Johnson, "Oral Contraceptives Decrease the Prevalence of Ovarian Cancer in the Hen," *Cancer Prevention Research* 5, no. 2 (February 2012): 343–349, doi:10.1158/1940-6207.capr-11-0344; Kristine Ansenberger et al., "Decreased Severity of Ovarian Cancer and Increased Survival in Hens Fed a Flaxseed-Enriched Diet for 1 Year," *Gynecologic Oncology* 117, no. 2 (May 2010): 341–347, doi:10.1016/j.ygyno.2010.01.021; Hawkridge, "Chicken Model"; Ana D. Bernardo, Sólveig Thorsteindóttir, and Christine L. Mummery, "Advantages of the Avian Model for Human Ovarian Cancer," *Molecular and Clinical Oncology* 3, no. 6 (2015): 1191–1198, doi:10.3892/mco.2015.619; Donna K. Carver et al., "Reduction of Ovarian and Oviductal Cancers in Calorie-Restricted Laying Chickens," *Cancer Prevention Research* 4, no. 4 (April 2011): 562–567, doi:10.1158/1940-6207.capr-10-0294.

6.   Hawkridge, "Chicken Model"; "EGFR," *NCI Dictionary of Cancer Terms* (Bethesda, MD: National Cancer Institute, n.d.), accessed March 3, 2022, https://www.cancer.gov/publications/dictionaries/cancer-terms/def/egfr; "Breast Cancer HER2 Status," American Cancer Society, accessed March 3, 2022, https://www.cancer.org/cancer/breast-cancer/understanding-a-breast-cancer-diagnosis/breast-cancer-her2-status.html; Shannon K. Laughlin-Tommaso, "CA 125 Test: A Screening Test for Ovarian Cancer?" Mayo Clinic, last modified August 6, 2020, https://www.mayoclinic.org/diseases-conditions/ovarian-cancer/expert-answers/ca-125/faq-20058528; Karen H. Vousden and David P. Lane, "p53 in Health and Disease," *Nature Reviews Molecular Cell Biology* 8, no. 4 (2007): 275–283, doi:10.1038/nrm2147; Whasun Lim et al., "SERPINB3 in the Chicken Model of Ovarian Cancer: A Prognostic Factor for Platinum Resistance and Survival in Patients with Epithelial Ovarian Cancer," *PLoS ONE* 7, no. 11 (November 21, 2012), doi:10.1371/journal.pone.0049869; Laurent Brard, email correspondence, May 29, 2022.

7.   "BRCA Gene Mutations: Cancer Risk and Genetic Testing Fact Sheet," National Cancer Institute, accessed March 3, 2022, https://www.cancer.gov/about-cancer/causes-prevention/genetics/brca-fact-sheet; "10 Things to Know about BRCA Genes," Texas Oncology, accessed March 3, 2022, https://www.texasoncology.com/services-and-treatments/genetic-testing/things-to-know-about-brca-genes; "BRCA Gene Mutations," National Cancer Institute.

8.   "Questions about the BRCA1 and BRCA2 Gene Study and Breast Cancer," Genome.gov, last modified June 1, 2012, https://www.genome.gov/10000940/brca1brca2-study-faq#al-3; Kelsey Lewis et al., "Recommendations and Choices for BRCA Mutation Carriers at Risk for

Ovarian Cancer: A Complicated Decision," *Cancers* 10, no. 2 (February 2018): 57, doi:10.3390/cancers10020057.

9. William H. Parker et al., "Long-Term Mortality Associated with Oophorectomy Compared with Ovarian Conservation in the Nurses' Health Study," *Obstetrics and Gynecology* 121, no. 4 (April 2013): 709–716, doi:10.1097/aog.0b013e3182864350; Radina Eshtiaghi, Alireza Esteghamati, and Manouchehr Nakhjavani, "Menopause Is an Independent Predictor of Metabolic Syndrome in Iranian Women," *Maturitas* 65, no. 3 (March 2010): 262–266, doi:10.1016/j.maturitas.2009.11.004; "Premature and Early Menopause: Causes, Diagnosis, and Treatment," Cleveland Clinic, last modified October 22, 2019, https://my.clevelandclinic.org/health/diseases/21138-premature-and-early-menopause; Hannaford Edwards et al., "The Many Menopauses: Searching the Cognitive Research Literature for Menopause Types," *Menopause* 26, no. 1 (January 2019): 45–65, doi:10.1097/gme.0000000000001171; Madison A. Price et al., "Early and Surgical Menopause Associated with Higher Framingham Risk Scores for Cardiovascular Disease in the Canadian Longitudinal Study on Aging," *Menopause* 28, no. 5 (May 2021): 484–490, doi:10.1097/gme.0000000000001729; Karen M. Tuesley et al., "Hysterectomy with and without Oophorectomy and All-Cause and Cause-Specific Mortality," *American Journal of Obstetrics and Gynecology* 223, no. 5 (November 2020): xx, doi:10.1016/j.ajog.2020.04.037; Lewis, "Recommendations and Choices for BRCA."

10. Jie Cui, Yong Shen, and Rena Li, "Estrogen Synthesis and Signaling Pathways during Aging: From Periphery to Brain," *Trends in Molecular Medicine* 19, no. 3 (March 2013): 197–209, doi:10.1016/j.molmed.2012.12.007; Radwa Barakat et al., "Extra-Gonadal Sites of Estrogen Biosynthesis and Function," *BMB Reports* 49, no. 9 (September 2016): 488–496, doi:10.5483/bmbrep.2016.49.9.141; "What Is Menopause?" National Institute on Aging, last modified September 30, 2021, https://www.nia.nih.gov/health/what-menopause; "Perimenopause: What Is It, Symptoms and Treatment," Cleveland Clinic, last modified October 5, 2021, https://my.clevelandclinic.org/health/diseases/21608-perimenopause.

11. Ding-Hao Liu et al., "Age-Related Increases in Benign Paroxysmal Positional Vertigo Are Reversed in Women Taking Estrogen Replacement Therapy: A Population-Based Study in Taiwan," *Frontiers in Aging Neuroscience* 9 (December 12, 2017), doi:10.3389/fnagi.2017.00404; Fernando Lizcano and Guillermo Guzmán, "Estrogen Deficiency and the Origin of Obesity during Menopause," *BioMed Research International* 2014 (March 2014): 1–11, doi:10.1155/2014/757461; Michael C. Honigberg et al., "Association of Premature Natural and Surgical Menopause with Incident Cardiovascular Disease," *Journal of the American Medical Association* 322, no. 24 (December 2019): 2411,

doi:10.1001/jama.2019.19191; D. Pu et al., "Metabolic Syndrome in Menopause and Associated Factors: A Meta-analysis," *Climacteric* 20, no. 6 (October 2017): 583–591, doi:10.1080/13697137.2017.1386649.

12. Maunil K. Desai and Roberta D. Brinton, "Autoimmune Disease in Women: Endocrine Transition and Risk across the Lifespan," *Frontiers in Endocrinology* 10 (April 29, 2019), doi:10.3389/fendo.2019.00265; Salman Assad et al., "Role of Sex Hormone Levels and Psychological Stress in the Pathogenesis of Autoimmune Diseases," *Cureus* 9, no. 6 (June 5, 2017), doi:10.7759/cureus.1315.

13. Jessica Grose, "When Your Home Is a Hormonal Hellscape," *New York Times*, May 27, 2021, https://www.nytimes.com/2021/05/26/parenting/menopause-perimenopause-puberty.html.

14. "Menopause: Diagnosis and Treatment," Mayo Clinic, last modified October 14, 2020, https://www.mayoclinic.org/diseases-conditions/menopause/diagnosis-treatment/drc-20353401; "Staying Healthy after Menopause," Johns Hopkins Medicine, accessed March 3, 2022, https://www.hopkinsmedicine.org/health/conditions-and-diseases/staying-healthy-after-menopause; Evrim Cakir et al., "Comparison of the Effects of Surgical and Natural Menopause on Epicardial Fat Thickness and γ-Glutamyltransferase Level," *Menopause* 18, no. 8 (August 2011): 901–905, doi:10.1097/gme.0b013e31820ca95e.

15. Remy Melina, "Sex-Change Chicken: Gertie the Hen Becomes Bertie the Cockerel," Live Science, last modified March 31, 2011, https://www.livescience.com/13514-sex-change-chicken-gertie-hen-bertie-cockerel.html; J. Pitino, "Spontaneous Sex Reversal: Is That My Hen Crowing?!" *Backyard Poultry,* last modified June 21, 2021, https://backyardpoultry.iamcountryside.com/feed-health/spontaneous-sex-reversal-is-that-my-hen-crowing/; "UCP Episode 018: Spontaneous Sex Reversal in Chickens—My Hen Just Became a Rooster!" podcast audio, July 17, 2013, http://www.urbanchickenpodcast.com/ucp-episode-018/; Jacquie Jacob, "Sex Reversal in Chickens," Small and Backyard Poultry—Welcome to the Poultry Extension Website, accessed March 3, 2022, https://poultry.extension.org/articles/poultry-anatomy/avian-reproductive-female/sex-reversal-in-chickens-kept-in-small-and-backyard-flocks/; Ker Than, "Half-Male Chicken Mystery Solved," *National Geographic*, last modified March 18, 2010, https://www.nationalgeographic.com/culture/article/100315-half-male-half-female-chickens.

16. Masha Gessen, "Masha Gessen on the Ins and Outs of Russia (Ep. 73)," Tyler Cowen, podcast audio, August 14, 2019, https://conversationswithtyler.com/episodes/masha-gessen/.

17. Jen Gunter, "Women Can Have a Better Menopause: Here's How," *New York Times*, May 26, 2021, https://www.nytimes.com/2021/05/25/opinion/feminist-menopause.html; Chris Harris, "Finding the Value in

Processing Spent Laying Hens," Poultry Site, December 20, 2019, https://www.thepoultrysite.com/articles/finding-the-value-in-processing-spent-laying-hens.

18. "Killer Whale," NOAA Fisheries, accessed March 3, 2022, https://www.fisheries.noaa.gov/species/killer-whale; Stuart Nattrass et al., "Postreproductive Killer Whale Grandmothers Improve the Survival of Their Grandoffspring," *Proceedings of the National Academy of Sciences* 116, no. 52 (December 2019): 26669–266673, doi:10.1073/pnas.1903844116; Daryl P. Shanley et al., "Testing Evolutionary Theories of Menopause," *Proceedings of the Royal Society B: Biological Sciences* 274, no. 1628 (September 2007): 2943–2949, doi:10.1098/rspb.2007.1028; Gail A. Greendale et al., "Changes in Regional Fat Distribution and Anthropometric Measures across the Menopause Transition," *Journal of Clinical Endocrinology and Metabolism* 106, no. 9 (August 2021): 2520–2534, doi:10.1210/clinem/dgab389; Chloe Shantz-Hilkes, "Jen Gunter Says Menopause Is a Heck of a Lot Less Scary When We Talk about It," CBC, May 27, 2021, https://www.cbc.ca/radio/asithappens/as-it-happens-the-thursday-edition-1.6042622/jen-gunter-says-menopause-is-a-heck-of-a-lot-less-scary-when-we-talk-about-it-1.6042625; Tove Danovich, "America Stress-Bought All the Baby Chickens," *New York Times*, March 28, 2020, https://www.nytimes.com/2020/03/28/style/chicken-eggs-coronavirus.html.

INDEX